Lecture Notes in Mathematics   1618

Editors:
A. Dold, Heidelberg
F. Takens, Groningen

Springer
Berlin
Heidelberg
New York
Barcelona
Budapest
Hong Kong
London
Milan
Paris
Santa Clara
Singapore
Tokyo

Gilles Pisier

Similarity Problems and Completely Bounded Maps

Springer

Author

Gilles Pisier
Mathematics Department
Texas A&M University
College Station, TX 77843, USA

and

Equipe d'Analyse
Université Paris VI
Case 186, 4 Place Jussieu
F-75252 Paris Cedex 05, France
E-mail: gip@ccr.jussieu.fr

Cataloging-in-Publication Data applied for

Die Deutsche Bibliothek - CIP-Einheitsaufnahme

Pisier, Gilles:
Similarity problems and completely bounded maps / Gilles
Pisier. - Berlin ; Heidelberg ; New York ; Barcelona ; Budapest
; Hong Kong ; London ; Milan ; Paris ; Tokyo : Springer, 1995
  (Lecture notes in mathematics ; 1618)
  ISBN 3-540-60322-0
NE: GT

Mathematics Subject Classification (1991):
Primary: 47A05, 46L05, 43A65
Secondary: 47A20, 47B10, 42B30, 46E40, 46L57, 47C15, 47D50

ISBN 3-540-60322-0 Springer-Verlag Berlin Heidelberg New York

© Springer-Verlag Berlin Heidelberg 1996
Printed in Germany

Typesetting: Camera-ready TeX output by the author
SPIN: 10479463   46/3142-543210 - Printed on acid-free paper

Foreword

These notes revolve around three similarity problems, appearing in three different contexts, but all dealing with the space $B(H)$ of all bounded operators on a complex Hilbert space H. The first one deals with group representations, the second one with C^*-algebras and the third one with the disc algebra. We describe them in detail in the introduction which follows.

This volume is devoted to the background necessary to understand these three *open* problems, to the solutions that are known in some special cases and to numerous related concepts, results, counterexamples or extensions which their investigation has generated.

For instance, we are naturally lead to study various Banach spaces formed by the matrix coefficients of group representations. Furthermore, we discuss the closely connected Schur multipliers and Grothendieck's striking characterization of those which act boundedly on $B(H)$.

While the three problems seem different, it is possible to place them in a common framework using the key concept of *"complete boundedness"*, which we present in detail. In some sense, completely bounded maps can also be viewed as spaces of "coefficients" of C^*-algebraic representations, if we allow "$B(H)$-valued coefficients", this is the content of the fundamental factorization property of these maps, which plays a central rôle in this volume.

Using this notion, the three problems can all be formulated as asking whether "boundedness" implies "complete boundedness" for linear maps satisfying certain additional algebraic identities.

Finally, a word on the structure of this book: this is definitely a lecture notes volume. Each chapter corresponds roughly to a lecture. In each one, we try to reach quickly some main point, without too many side remarks. This usually corresponds to the actual lecture. Once this point has been made, we then allow ourselves to develop all sorts of additional comments and a guide to the literature (the "notes and remarks") which expand on the first part, and which (for lack of time usually) the audience of the lecture is invited to read.

The main body of the notes is essentially self contained and can be read by anyone familiar with basic Functional and Harmonic Analysis, as presented for example in Rudin's books [R2, R4]. We believe this volume may be used as the basis for an advanced graduate course in Functional Analysis.

These notes are originally based on a course given during a summer school organized by S. Negrepontis in Spetses (Greece) in july 92, and on courses given in Texas A&M and in Paris VI in 92/93. I am very grateful to all those who subsequently provided me with more information, additional references and corrected misprints or errors of all kind. In particular, it is a pleasure to thank M. Bożejko, A. Hess, J. Holbrook, C. Le Merdy, V. Mascioni, V. Paulsen, V. Peller, G. Popescu, F. Wattbled and Q. Xu.

The research for this work was partially supported by the N. S. F..

In conclusion, I would like to thank Robin Campbell, who typed most of the book, for her outstanding job.

Table of Contents

0. Introduction. Description of contents

We will discuss three "similarity problems", each of which is still an open problem at least in some form.

Historically, the first problem arose in the context of representations of groups. Let H be a Hilbert space (over the complex field). Let G be a group and let $\pi\colon G \to B(H)$ be a *representation* of G. This means that

$$\pi(1) = I \quad \text{and} \quad \pi(st) = \pi(s)\pi(t) \qquad \forall s, t \in G.$$

Hence necessarily $\pi(t)$ is invertible for all t in G and $\pi(t^{-1}) = \pi(t)^{-1}$. If $\pi(t)$ is a unitary operator (i.e. if $\pi(t^{-1}) = \pi(t)^*$) for all t in G, we say that π is a *unitary representation*. If G is a locally compact group, we will say that π is continuous if π is continuous when $B(H)$ is equipped with the strong operator topology. (When G is discrete, of course every representation is continuous.)

Problem 0.1. *Let $\pi\colon G \to B(H)$ be a continuous representation of a locally compact group G. Assume π uniformly bounded, i.e. assume*

$$\sup_{t \in G} \|\pi(t)\|_{B(H)} < \infty.$$

Does there exist an invertible operator $S\colon H \to H$ such that the representation

$$\tilde{\pi}(t) = S^{-1}\pi(t)S$$

is a unitary representation?

When this holds, we will say that π is unitarizable. In other words, is every uniformly bounded continuous representation of G unitarizable?

Notes: Let $u\colon G \to B(H)$ be a unitary representation. Then for any invertible operator $S\colon H \to H$ the formula

$$\tilde{u}(t) = Su(t)S^{-1}$$

defines a uniformly bounded representation with

$$\sup_{t \in G} \|\tilde{u}(t)\| = C < \infty$$

for some $C \le \|S\|\,\|S^{-1}\|$. In the discrete case, the preceding problem asks whether all uniformly bounded representations can be obtained in this way.

This problem remained open for a while until, in 1955, Ehrenpreis and Mautner [EM] (see also [KZ]) gave a counterexample for $G = SL_2(\mathbb{R})$. Later it was realized that rather simpler counterexamples can be described on the free groups with at least 2 generators (cf. e.g. [MZ1-2, FTP, PyS, B5, W2]). In the positive direction, the most general result seems to be Dixmier's theorem (1955) which says that if G is *amenable* then the answer to Problem 0.1 is affirmative (see Theorem 0.6 below), but the converse remains an open problem. Thus we have

Revised Problem 0.1. *If all uniformly bounded continuous representations of a group G are unitarizable, is G necessarily amenable?*

For more on all this see Theorem 0.6 and chapter 2 below.

The second problem arose in the C^*-algebra context. Let A be a C^*-algebra. A linear operator $u\colon A \to B(H)$ will be called a homomorphism if

$$\forall x, y \in A \qquad u(xy) = u(x)u(y).$$

If A has a unit, we will usually also require $u(1) = I$. The following problem was raised by Kadison [K1].

Problem 0.2. *Let $u\colon A \to B(H)$ be a bounded unital homomorphism on a C^*-algebra A. Does there exist an invertible operator $S\colon H \to H$ such that the map*

$$\tilde{u}(x) = S^{-1}u(x)S$$

is a $*$-representation of A i.e. is such that

$$\tilde{u}(x^*) = \tilde{u}(x)^*?$$

In other words, is every bounded homomorphism u from A into $B(H)$ similar to a $*$-representation?

When this holds, Kadison [K1] calls u orthogonalizable.
Notes: (i) Let $\rho\colon A \to B(H)$ be a $*$-representation. Then necessarily $\|\rho\| \leq 1$. This is checked as follows. Assume A has a unit. Clearly ρ maps unitaries to unitaries. Hence $\|\rho(u)\| \leq 1$ for any unitary u. Then any hermitian in the unit ball is the real part of a unitary. Hence $\|\rho(u)\| \leq 1$ for any hermitian in the unit ball. Finally, since $\|x^*x\| = \|x\|^2$ in a C^*-algebra, we conclude that $\|\rho(x)\|^2 = \|\rho(x)^*\rho(x)\| = \|\rho(x^*x)\| \leq \|x^*x\| = \|x\|^2$, so that $\|\rho\| \leq 1$. We recall in passing the Russo-Dye Theorem, which gives an alternate proof of this: if A has a unit, the unit ball of A is the closed convex hull of the unitaries of A.
(ii) Let $\rho\colon A \to B(H)$ be a $*$-representation. Then for any invertible operator $S\colon H \to H$ the formula

$$u(x) = S\rho(x)S^{-1}$$

defines a homomorphism on A into $B(H)$ and we have

$$\|u\| \leq C < \infty$$

for some $C \leq \|S\|\,\|S^{-1}\|$. The above problem 0.2 asks precisely whether all bounded homomorphisms are of this form.

In full generality problem 0.2 is still open although important partial results were obtained by E. Christensen [C3] and U. Haagerup [H1]. In particular, Haagerup [H1] proved that if in addition u is cyclic (i.e. there is an element ξ in H, such that $\overline{u(A)\xi} = H$, such a ξ is called a cyclic vector) then the answer to problem 0.2 is affirmative. See chapter 7 for more details on this.

The preceding problem is closely related to an other famous open problem, the so-called derivation problem. Recall that, if A is a subalgebra of an algebra B, a derivation is a linear map $\delta\colon A \to B$ such that $\delta(ab) = \delta(a)b + a\delta(b)$. For example, given T in B, the map $\delta_T\colon A \to B$, defined by $\delta_T(a) = aT - Ta$, is a derivation. Such derivations are called inner. Now we can state the well known derivation problem (see chapter 4 for more on this).

Problem 0.2'. *Let $A \subset B(H)$ be a C^*-subalgebra. Is every derivation $\delta\colon A \to B(H)$ inner?*

This problem can also be formulated as the vanishing of the first Hochschild cohomology group $H^1(A, B(H))$ of a C^*-algebra A (with "coefficients in $B(H)$"), see [J, Hel, SS] for more in this direction. Many variants of this problem have a positive solution, for instance any derivation $\delta\colon A \to B(H)$ with range included into A is inner.

The third similarity problem is probably the most famous one, it concerns "polynomially bounded operators". Before we give the precise definition, recall the following classical inequality of von Neumann satisfied by any contraction T on a Hilbert space H.

Von Neumann's inequality. *For any T in $B(H)$ with $\|T\| \le 1$ and for any polynomial P we have*

$$(vN) \qquad\qquad \|P(T)\| \le \|P\|_\infty$$

where $\|P\|_\infty$ is defined by

$$\|P\|_\infty = \sup\{|P(z)| \mid z \in \mathbb{C}, |z| = 1\}.$$

For a proof (actually several ones) see chapter 1 below.
Let $D = \{z \in \mathbb{C} \mid |z| < 1\}$ and

$$\partial D = \mathbf{T} = \{z \in \mathbb{C} \mid |z| = 1\}.$$

By the maximum principle, we clearly have

$$(0.1) \qquad\qquad \|P\|_\infty = \sup_{z \in D} |P(z)| = \sup_{z \in \overline{D}} |P(z)|,$$

and actually this holds for any continuous function $P\colon \overline{D} \to \mathbb{C}$ which is analytic in the interior (i.e. on D).

Definition. *We will say that an operator* $T\colon H \to H$ *is polynomially bounded (in short p.b.) if there is a constant* C *such that for all polynomials* P *we have*

$$\|P(T)\| \le C\|P\|_\infty.$$

In that case, we will say that T *is p.b. with constant* C.

We can now state the third version (this one purely operator theoretic) of the similarity problem.

Problem 0.3. *Let* $T\colon H \to H$ *be a polynomially bounded operator. Does there exist an invertible operator* $S\colon H \to H$ *such that*

$$\|S^{-1}TS\| \le 1?$$

Usually, an invertible operator $S\colon H \to H$ is called a *similarity* and two operators T_1, T_2 on H are called *similar* if there is a similarity S such that $T_1 S = S T_2$ or equivalently $T_2 = S^{-1} T_1 S$.
Note: Let T be a contraction (i.e. such that $\|T\| \le 1$) on H and let $S\colon H \to H$ be invertible. Let $\widetilde{T} = STS^{-1}$. Then clearly $\widetilde{T}$ is polynomially bounded.
Indeed for any polynomial P, $P(\widetilde{T}) = SP(T)S^{-1}$ hence by (vN)

$$\|P(\widetilde{T})\| \le \|S\|\,\|S^{-1}\|\,\|P(T)\|$$
$$\le C\|P\|_\infty$$

with $C = \|S\|\,\|S^{-1}\|$. The preceding problem asks precisely whether all polynomially bounded operators arise in this way.

Problem 0.3 was presented by Halmos in [Ha1], and is often referred to as "the Halmos problem". Actually the original question (due to Sz.-Nagy) was: if T is a power bounded operator, i.e. if $\sup_{n \ge 0} \|T^n\| < \infty$, is T similar to a contraction? This was quickly disproved by Foguel [Fo] but then problem 0.3 emerged as the revised version of this question, and it has resisted since then. In chapter 2 below we will give examples of power bounded operators which are not similar to contractions, following [B2]. Note that if T is invertible, then T and T^{-1} are both power bounded iff T is similar to a unitary operator. This is one of the original results of Sz.-Nagy at the root of these questions, but of course it follows from Dixmier's above mentioned theorem: since $\mathbb{Z}$ is amenable, and $n \to T^n$ is a representation of $\mathbb{Z}$, this representation is uniformly bounded iff it is unitarizable.

Let $A(D)$ be the disc algebra, *i.e.* the completion of the space of polynomials for the norm $\|\ \|_\infty$. Clearly, an operator T is polynomially bounded iff the map $P \to P(T)$ extends to a bounded homomorphism on $A(D)$. Since the latter algebra is not self-adjoint, its structure is quite different from that of a C^*-algebra. Hence Problems 0.3 and 0.2 appear quite different a priori. Nevertheless, we will study them in chapter 4 in a common framework using the notion of complete boundedness.

The preceding three problems can all be viewed as renorming problems. Consider Problem 0.3 first. The following fact is a simple observation.

Proposition 0.4. *Let $T: H \to H$ be an operator.*

(i) *The operator T is polynomially bounded iff there is an equivalent norm on H for which the operator T is polynomially bounded with constant 1.*

(ii) *The operator T is similar to a contraction iff there is an equivalent Hilbertian norm for which T is a contraction.*

(iii) *The operator T is power bounded iff there is an equivalent norm on H for which T is a contraction.*

Proof: This is an easy exercise left to the reader.

Note that if T is polynomially bounded (resp. power bounded) then the formula

$$|||x||| = \sup\{\|P(T)x\| \mid \|P\|_\infty \le 1\}$$

(resp. $|||x||| = \sup\{\|T^n x\| \mid n \ge 0\}$) defines an equivalent norm on H such that $|||P(T)x||| \le \|P\|_\infty |||x|||$ (resp. $|||T^n x||| \le |||x|||$) so that T satisfies von Neumann's inequality (resp. is a contraction) with respect to the new norm. However, the space H equipped with the norm $||| \ |||$ is only a *Banach* space. Let us denote it by X. Obviously the identity on H induces an invertible operator $S: H \to X$ such that the operator $\tilde{T} = STS^{-1}: X \to X$ satisfies (vN) (resp. $\|\tilde{T}\| \le 1$) but this is on the *Banach* space X. To solve Problem 0.3, we need to find an equivalent *Hilbertian* norm on H for which T becomes a contraction.

Equivalently, Problem 0.3 can be reformulated as follows. Let $T: X \to X$ be an operator on a Banach space satisfying (vN) i.e. such that $\|P(T)\| \le \|P\|_\infty$ for all polynomial P. If X admits an equivalent Hilbertian norm, is there an equivalent Hilbertian norm which preserves the property (vN) for the operator T?

Note that the other problems can be rephrased in the same way. In particular Problem 0.2 also can be viewed as a renorming problem. Consider a homomorphism $u: A \to B(H)$, assume that A has a unit and $u(1) = I$. Observe that u is a *-representation iff it is a contractive homomorphism. Indeed, as already mentioned above (cf. the notes following Problem 0.2) any *-representation is contractive and conversely if u is contractive it maps unitaries to unitaries (since unitary operators are nothing but invertible contractions with a contractive inverse), hence we have $u(x)^* = u(x^*)$ for all unitary operators x in A and since these generate A, u is a *-representation.

Then u is bounded (resp. similar to a *-representation) iff there is an equivalent norm (resp. Hilbertian norm) on H for which u becomes a contractive homomorphism. This can be proved exactly as Proposition 0.4 above but replacing the algebra of all polynomials equipped with the norm (0.1) by a C^*-algebra.

A similar discussion applies in the group case. Consider a uniformly boun-ded representation $\pi: G \to B(H)$ on a group G. Then π is unitary iff $\sup_{t \in G} \|\pi(t)\| \le 1$. Indeed, if $\|\pi(t)\| \le 1$ for all t then since $\pi(t)^{-1} = \pi(t^{-1})$, $\pi(t)$ is an invertible isometry hence a unitary operator. Therefore similar comments apply in this case. A representation $\pi: G \to B(H)$ is uniformly bounded (resp. unitarizable) iff there is an equivalent norm (resp. Hilbertian norm) on H for which π becomes uniformly bounded by 1. To illustrate this, let us discuss *amenable* groups.

Definition 0.5. *A locally compact group G is called amenable if there exists a left invariant mean on G, i.e. if there exists a positive linear functional $\varphi: L_\infty(G) \to \mathbb{C}$ satisfying $\|\varphi\| = \varphi(1) = 1$, and*

$$\forall f \in L_\infty(G) \quad \forall t \in G \qquad \varphi(\delta_t * f) = \varphi(f).$$

(Recall that $\delta_t * f(s) = f(t^{-1}s) \ \forall s \in G$.)

We can now state Dixmier's theorem.

Theorem 0.6. *If G is amenable, every uniformly bounded continuous representation π on G is unitarizable, i.e. Problem 0.1 has an affirmative answer in the amenable case. More precisely, there is an invertible operator $S : H \to H$, with $\|S\|\|S^{-1}\| \leq \sup_{t \in G} \|\pi(t)\|^2$, such that $S^{-1}\pi(.)S$ is a unitary representation.*

Proof: Assume $\sup_{t \in G} \|\pi(t)\| = C < \infty$. Then for any x, y in H, we denote

$$\forall t \in G \qquad f_{xy}(t) = \langle \pi(t^{-1})x, \pi(t^{-1})y \rangle.$$

Observe that $f_{xy} \in L_\infty(G)$. Let φ be an invariant mean on G as defined above. We define

$$\||x\|| = (\varphi(f_{xx}))^{1/2}.$$

We have clearly $f_{xx} \leq C^2\|x\|^2$ hence

$$\||x\|| \leq C\|x\|.$$

On the other hand $\|x\|^2 \leq \|\pi(t)\|^2 f_{xx}(t) \leq C^2 f_{xx}(t)$, hence $\|x\| \leq C\||x\||$. Clearly, $\|| \ \||$ is a *Hilbertian* norm on H since it can be derived from the scalar product $\ll x, y \gg = \varphi(f_{xy})$. Finally, if we equip H with the new norm $\|| \ \||$, by the invariance of φ, π becomes unitary. Indeed we have clearly for all s in G

$$(0.2) \qquad \||\pi(s)x\||^2 = \varphi(\delta_s * f_{xx}) = \varphi(f_{xx}) = \||x\||^2$$

which shows that $\pi(s)$ becomes unitary. Finally, since $(H, \|| \ \||)$ is Hilbertian and isomorphic to H, it is actually isometric to $(H, \| \ \|)$. Hence there is an invertible linear mapping $S: H \to H$ such that

$$\forall x \in H \qquad \||Sx\|| = \|x\| \text{ or equivalently } \||x\|| = \|S^{-1}x\|.$$

Viewing S as an operator from $(H, \| \ \|)$ to $(H, \| \ \|)$, we obtain $\|S\| \leq C$, $\|S^{-1}\| \leq C$ and by (0.2), $t \to S^{-1}\pi(t)S$ is a unitary representation. $\square$

Among the examples of amenable groups are all the compact groups (then the Haar integral is an invariant mean) and all the Abelian groups. Let us show for instance that $\mathbb{Z}$ is amenable. Note that if G is discrete $L_\infty(G) = \ell_\infty(G)$. Let $\mathcal{U}$ be a nontrivial ultrafilter on $\mathbb{N}$ (or a so-called Banach limit). Let $f \in \ell_\infty(G)$. We define

$$\varphi(f) = \lim_{\substack{n \to \infty \\ \mathcal{U}}} \left[\frac{1}{2n+1} \sum_{k=-n}^{k=+n} f(k) \right].$$

Clearly φ is positive moreover $\varphi(1) = 1 = \|\varphi\|$ and for each fixed t in $\mathbb{Z}$, the difference

$$\frac{1}{2n+1} \sum_{k=-n}^{k=+n} (f(t+k) - f(k))$$

tends to zero when $n \to \infty$. Hence φ is an invariant mean on $\mathbb{Z}$ so that $\mathbb{Z}$ is amenable.

The free group on two generators say a, b is denoted by $\mathbb{F}_2$. It is the typical example of a nonamenable group. To check this, assume there is an invariant mean φ on $\mathbb{F}_2$. We will reach a contradiction. Since this group is infinite, we have necessarily $\varphi(1_{\{t\}}) = 0$ for all t. Moreover, we have a disjoint partition

$$\mathbb{F}_2 = F(a) \cup F(a^{-1}) \cup F(b) \cup F(b^{-1}) \cup \{e\}$$

where $F(x)$ is the set of (reduced) words which have x as their first letter. Hence this decomposition implies

$$1 = \varphi(1_{\mathbb{F}_2}) = \varphi(1_{F(a)}) + \varphi(1_{F(a^{-1})}) + \varphi(1_{F(b)}) + \varphi(1_{F(b^{-1})}).$$

On the other hand we have

$$F(a) = a[\mathbb{F}_2 - F(a^{-1})]$$
$$F(a^{-1}) = a^{-1}[\mathbb{F}_2 - F(a)]$$
$$F(b) = b[\mathbb{F}_2 - F(b^{-1})]$$
$$F(b^{-1}) = b^{-1}[\mathbb{F}_2 - F(b)],$$

hence

$$\varphi(1_{F(a)}) = 1 - \varphi(1_{F(a^{-1})})$$
$$\varphi(1_{F(a^{-1})}) = 1 - \varphi(1_{F(a)})$$
$$\varphi(1_{F(b)}) = 1 - \varphi(1_{F(b^{-1})})$$
$$\varphi(1_{F(b^{-1})}) = 1 - \varphi(1_{F(b)}).$$

If we add the last four identities we obtain $1 = 4 - 1$ which is the desired contradiction. $\square$

We will now briefly describe the contents of this volume, chapter by chapter. Note that each chapter starts with a brief summary and is followed by a set of "Notes and Remarks". The reader should always consult them for more information, especially for more precise references and credits.

In chapter 1, we give at least three different proofs of von Neumann's inequality and one of Ando's inequality. We also prove Sz.-Nagy's and Ando's dilation theorems which say respectively that in the cases $n = 1$ and $n = 2$ every n-tuple of mutually commuting contractions can be dilated into an n-tuple of mutually commuting unitary operators (see Theorems 1.1 and 1.2). We also discuss the case of n mutually commuting contractions with $n > 2$, for which counterexamples are known (see Proposition 1.6 and the remarks before it), although numerous questions remain open.

We introduce the key notion of semi-invariance in Sarason's sense (see Theorem 1.7) and we explain why Hilbert spaces are the only Banach spaces on which von Neumann's inequality can hold (see Theorem 1.9).

In the Notes and Remarks following chapter 1, we describe various extensions of von Neumann's inequality in particular for polynomials in several (possibly non-commuting) variables.

In chapter 2, we construct examples of non-unitarizable uniformly bounded representations $\pi\colon G \to B(H)$ when G is the free group $\mathbb{F}_N$ with $N \geq 2$ generators. The approach we follow (due to Bożejko and Fendler) leads rather rapidly to such an example (Theorem 2.1, Lemma 2.2, Corollary 2.3). In addition, we briefly discuss amenable groups, in particular we give the well known Kesten-Hulanicki criterion (Theorem 2.4) of amenability and we connect it to our approach (Theorem 2.5, Lemma 2.7). Actually, this approach also works in the case when G is merely a semi-group with unit embeddable into a group, for instance when $G = \mathbb{N}$. In the latter case, this yields (Corollary 2.14) examples of power bounded operators on Hilbert space which are not polynomially bounded. Just like in the first construction of such examples in 1964 (due to Foguel), we rely heavily on the specific properties of Hadamard lacunary sequences in Fourier analysis (see Lemma 2.10). We also discuss briefly analogous properties for H^∞-functions with values in a von Neumann algebra (Remark 2.11) and we propose a conjecture in that setting. In the appendix, we include a proof of Schur's criterion for boundedness of a matrix on ℓ_2 and related facts.

In the notes and remarks on chapter 2, we make a special effort to describe the various classes of functions defined on groups which are often considered in Harmonic Analysis and we relate them to the spaces $B(G)$ (resp. $B_c(G)$) of coefficients of unitary (resp. uniformly bounded by c) representations on G.

In chapter 3, we apply the Hahn-Banach Theorem to give a criterion (see Theorem 3.4) for an operator $T\colon X \to Y$ (between Banach spaces) to factor through a Hilbert space. We include various properties of the resulting class – denoted by $\Gamma_2(X,Y)$ – of all such operators. One can use the same strategy (suitably adapted) for families $(T_i)_{i \in I}$ with $T_i\colon X \to Y$. This is the idea we use to treat the factorization properties of a completely bounded linear mapping $u\colon S \to B(X,Y)$ defined on a closed linear subspace $S \subset B(H)$. The basic factorization result is Theorem 3.6 which is the main result of this chapter. This statement implies both an extension theorem and a decomposition into completely positive maps (see Corollary 3.8). We quickly concentrate on the more classical (and more useful up to now) case when X and Y are both Hilbert spaces. We give various applications of the factorization of $c.b.$ maps, for instance we include (Corollary 3.12) a very simple description of the complete contractions $u\colon M_n \to M_n$. The latter corresponds to the particular case $H = X = Y = \ell_2^n$ and $S = B(H)$. We also show that the cb-norm of a map $u\colon S \to M_n$ can be computed by restricting oneself to $n \times n$ matrices with entries in S (see Proposition 3.13).

In addition, we show that the "matrix norm structure" of a C^*-algebra $A \subset B(H)$ (namely the sequence of normed spaces $(M_n(A))_{n \geq 1}$) is uniquely determined by the underlying C^*-algebra structure of A and we discuss the case when the range of $u\colon S \to B(H)$ lies inside a commutative C^*-subalgebra of $B(H)$ (see Corollary 3.18).

Finally, we give examples of maps u which are bounded but not $c.b.$. Such examples exist whenever S is infinite dimensional (see Remark 3.23). More precisely, for various choices of finite dimensional subspaces $S \subset B(H)$, we give concrete examples (Theorem 3.21) of maps $u\colon S \to B(H)$ with $\|u\|$ small and $\|u\|_{cb}$ large. These estimates show that whenever $\dim(S) \geq 5$, there is a map $u\colon S \to B(H)$ with $\|u\|_{cb} > \|u\|$ (see Remark 3.24).

In chapter 4, we study the notion of "*compression*" of a homomorphism $u\colon A \to B(X)$ defined on an algebra A with values in the algebra of bounded operators on a Banach space X. When X is Hilbertian, that notion is relative to the data of a *semi-invariant* subspace $E \subset X$, which means that $E = E_1 \ominus E_2$ for some invariant subspaces $E_2 \subset E_1$. Interestingly enough, it turns out that the notion of compression (relative to a pair $E_2 \subset E_1$ of invariant subspaces) makes sense when X is a Banach space. Actually, the basic underlying ideas are perhaps more transparent when presented in the Banach space context.

The main result of chapter 4 is Paulsen's Theorem showing that a unital homomorphism $u\colon A \to B(H)$ (here A is a unital subalgebra of $B(\mathcal{H})$) is $c.b.$ with $\|u\|_{cb} \leq C$ iff there is an invertible $S\colon H \to H$ with $\|S\|\,\|S^{-1}\| \leq C$ such that $a \to S^{-1}u(a)S$ is completely contractive (see Theorem 4.3). We deduce this result from a more general Banach space theoretic statement (Proposition 4.2) which describes the pairs of bounded homomorphisms $\pi\colon A \to B(Z)$ and $u\colon A \to B(X)$ such that u is similar to a compression of π (here A is any Banach algebra, Z, X arbitrary Banach spaces). In case A is a C^*-algebra and π is a $*$-homomorphism, semi-invariance implies invariance (indeed E invariant implies $E^{\perp}$ invariant), so that a compression is merely the restriction to an invariant subspace. Assuming furthermore that X, Z are Hilbert spaces, we obtain as a corollary (see Corollary 4.4) that a unital homomorphism $u\colon A \to B(H)$ is similar to a $*$-representation iff u is $c.b.$, and we deduce from this that a derivation $\delta\colon A \to B(H)$ is inner iff it is $c.b.$ (see Corollary 4.6). These results also have nice analogues in the non self-adjoint case: if we take for A the disc algebra, we obtain Paulsen's criterion that an operator $T\colon H \to H$ is similar to a contraction iff the associated homomorphism $u_T\colon A \to B(H)$ defined on polynomials by

$$u_T(f) = f(T)$$

is completely bounded. Moreover we have

(0.3) $\|u_T\|_{cb} = \inf\{\|S\|\,\|S^{-1}\| \mid \|S^{-1}TS\| \le 1\}.$

See Corollary 4.7. To some extent, this extends to operators $T\colon X \to X$ on a Banach space X (see Corollaries 4.13 and 4.14). Statements 4.8 and 4.9 are well known results closely related to Arveson's work on complete positivity, see [Pa1] for more background on their genesis.

To illustrate the use of c.b. maps, we derive from Corollary 4.7 some earlier classical facts: we show Rota's formula for the spectral radius

$$r(T) = \inf\{\|S^{-1}TS\| \mid S\colon H \to H \text{ invertible}\}$$

and deduce from it that a compact operator T with spectrum inside the closed unit disc is similar to a contraction (see Corollaries 4.10, 4.11 and 4.12).

The identity (0.3) allows to "quantify" the similarity problem. For instance, if T is an $n \times n$ matrix which is polynomially bounded with constant C, one can try to majorize $\|u_T\|_{cb}$ as a function of C and of the dimension n. Such an estimate (due to Bourgain) is stated as Theorem 4.15. Its proof is unfortunately technical and too difficult to be included in these notes, but we discuss some related, simpler but useful, inequalities (see Proposition 4.16 and Corollary 4.17). We also present some recent results (statements 4.18 to 4.21) which show that Bourgain's estimates can be extended to k-tuples of mutually commuting contractive (or polynomially bounded) matrices in M_n.

Finally, using the examples of bounded maps which are not c.b. given in chapter 3, we produce (Theorem 4.22) examples of (non-selfadjoint) unital subalgebras $A \subset B(H)$ admitting a contractive unital homomorphism $u\colon A \to B(H)$ which is not c.b. (see Theorem 4.22).

In chapter 5 we turn to the subject of Schur multipliers on $B(\ell_2)$, of which Grothendieck gave a striking characterization. The latter is the content of Theorem 5.1. Here we use the previous discussion of the class $\Gamma_2(X, Y)$ and show that Schur multipliers on $B(\ell_2)$ are in isometric correspondence with $\Gamma_2(\ell_1, \ell_\infty)$. We also show that a *bounded* Schur multiplier on $B(\ell_2)$ is *automatically completely bounded* (see Theorem 5.1).

In passing, we construct a tensor product of two Banach spaces E, F which is isometrically a predual of the space $\Gamma_2(E, F^*)$ (see Theorem 5.3) and we include Krivine's proof of Grothendieck's inequality which leads to the best known upper bound for the Grothendieck constant K_G (see Theorem 5.5). One of the classical reformulations of Grothendieck's inequality is in terms of polynomials of degree 2 in several complex variables (see Corollary 5.7). In statements 5.8 and 5.9 we show (following Varopoulos) that if the complex Grothendieck constant restricted to $n \times n$ -matrices is > 1, then there is a homogeneous polynomial of degree 2 in $2n$ variables for which the von Neumann inequality fails (for some $2n$ mutually commuting contractions).

Finally, we describe the extension of these results for Schur multipliers on $B(\ell_p)$ with $p \neq 2$ (see Theorems 5.10 and 5.11). The class $\Gamma_2(X,Y)$ is now replaced by the class of operators $\Gamma_p(X,Y)$ which factor through an L_p-space. We also consider the operators which factor through either a subspace, a quotient or a subspace of a quotient of an L_p-space. Here the structure of the proofs is very much the same as in the case $p = 2$, but the full details are left to the reader.

In the brief chapter 6, we concentrate on the Schur multipliers on $B(\ell_2)$ associated to a function $\varphi \colon \mathbb{N}^2 \to \mathbb{C}$ of the form

$$\varphi(i,j) = f(i+j) \qquad i,j \in \mathbb{N}.$$

Since the associated matrix $(\varphi(i,j))$ is a Hankel matrix, we call such multipliers "Hankelian". Clearly, any Hankelian Schur multiplier φ (acting on $B(\ell_2)$) leaves invariant the subspace of $B(\ell_2)$ formed of all Hankel matrices. We show that its restriction to that subspace is c.b. iff the Hankelian Schur multiplier itself is bounded ($= c.b.$) on the whole of $B(\ell_2)$. Moreover the corresponding norms are equal (see Theorem 6.1). There is an interesting application of this result to the class of those Fourier multipliers of the classical Hardy space H_1 which extend boundedly to the trace class valued H_1-space (see Theorem 6.2). This is connected with the recent theory of operator spaces where such multipliers are just c.b. multipliers on H_1 (see Remark 6.3 and the Notes and Remarks in Chapter 6).

Returning to the self-adjoint setting, we consider a discrete group G with its reduced C^*-algebra $C^*_\lambda(G)$. Let us say that a Schur multiplier φ on $B(\ell_2(G))$ is G-invariant if it is of the form $\varphi(s,t) = f(st^{-1})$ for some function $f \colon G \to \mathbb{C}$. Note that φ leaves $C^*_\lambda(G) \subset B(\ell_2(G))$ invariant. Here again we find (by analogous arguments) that a G-invariant Schur multiplier is bounded ($= c.b.$) on $B(\ell_2(G))$ iff its restriction to $C^*_\lambda(G)$ is c.b., with equal norms (see Theorem 6.4). Note that the Hankel theory corresponds to the replacement of G by the semi-group $\mathbb{N} \subset \mathbb{Z}$, exactly as in chapter 2.

In the Notes and Remarks after Chapter 6, we discuss at length a series of interesting problems raised by Peller [Pe1] on power bounded operators. Some of them are still open. We also explain there the notion of a completely bounded Schur multiplier on the Schatten class S_p and we propose a conjecture about them.

In chapter 7, the main result is Haagerup's solution of the similarity problem for cyclic homomorphisms $u \colon A \to B(H)$ on a C^*-algebra A (see Theorem 7.5). The proof we give is based on an inequality satisfied by *all* bounded linear mappings $u \colon A \to B(H)$ (see Theorem 7.1). This inequality is often referred to as the "non-commutative Grothendieck inequality", because it is proved using similar ideas as for Grothendieck's inequality. But this is misleading: indeed, if both the domain and the range of u are commutative C^*-algebras the result is essentially trivial (see Proposition 7.2), while Grothendieck's inequality remains non-trivial even in the commutative case. After the proof of Theorem 7.5, we include several variants. For instance, if a bounded homomorphism has a finite cyclic set, then it is automatically c.b.. Recall that it is an open problem (Problem 0.2) whether this remains true when u has an infinite cyclic set, or equivalently when u is nondegenerate, i.e. when we only assume that the set $\{u(a)\xi \mid a \in A, \ \xi \in H\}$ is dense in H.

The latter condition is of course automatic for a unital homomorphism. In the last part of Chapter 7, we give a positive answer to this question, but only for *certain* C^*-algebras A, namely for C^*-algebras without tracial states (such as $B(H)$ with dim $H = \infty$) or for simple, infinite C^*-algebras in Cuntz's sense (see Proposition 7.12 and Corollary 7.13). We also prove it for all nuclear C^*-algebras (Theorem 7.15). Finally we prove that for any given C^*-algebra A the similarity and the derivation problem are essentially equivalent (this was recently shown by Kirchberg).

In chapter 8, we introduce the notion of "p-complete boundedness" (in short p-c.b.) and we extend most of the results of chapters 3 and 4 to this new setting. The main difference is that we must distinguish between subspaces of L_p, quotients of L_p and subspaces of quotient (= quotients of subspaces) of L_p. These classes are all the same when $p = 2$, but they are different otherwise. In Theorem 8.1, we extend the fundamental factorization of c.b. maps to this new setting (and in Corollary 8.2, we record a consequence for the Schur multipliers of $B(\ell_p)$). In this case, the relevant analogue of a *-representation is a natural "action" on a quotient of a subspace of vector valued L_p. There also is a distinct statement for actions on a subspace of vector valued L_p (see Theorem 8.6), and another one for actions on vector valued L_p itself (see Corollary 8.7). Under suitable assumptions we also have an extension property of p-c.b. maps (see Corollary 8.8). Finally, we study the applications of these factorizations to the similarity problem for a unital homomorphism $u\colon A \to B(X)$ defined on a unital subalgebra $A \subset B(X_1)$, X, X_1 being Banach spaces. It turns out that u is p-c.b. iff there is a space Z which is a quotient of a subspace of (roughly) $L_p(X_1)$ and an isomorphism $\delta\colon X \to Z$ such that $a \to Su(a)S^{-1}$ is p-completely contractive.

In particular, when A is the disc algebra (embedded into $B(L_p(\mathbf{T}))$ in the natural way as multiplication operators) we extend Paulsen's criterion for similarity to a contraction (see Corollary 8.11).

1. Von Neumann's inequality and Ando's generalization

Summary: In this chapter, we prove (actually three times) von Neumann's inequality and its extension (due to Ando) for two mutually commuting contractions. We discuss the case of $n > 2$ mutually commuting contractions. We introduce the notion of semi-invariance. Finally, we show that Hilbert spaces are the only Banach spaces satisfying von Neumann's inequality.

First we will prove von Neumann's inequality (denoted by (vN) in the introduction) for an operator T on a finite dimensional Hilbert space. Equivalently we may identify T with an $n \times n$ matrix with complex entries and we compute the norms of the operators T or $P(T)$ as operators on the n-dimensional Hilbert space ℓ_2^n (i.e. $\mathbb{C}^n$ equipped with its standard Hilbert space structure).

Let us first assume T unitary. Then by diagonalization, there is a unitary operator v and $z_1, \ldots, z_n$ in ∂D such that

$$T = v^* \begin{pmatrix} z_1 & & 0 \\ & \ddots & \\ 0 & & z_n \end{pmatrix} v.$$

It follows that for any polynomial P

$$P(T) = v^* \begin{pmatrix} P(z_1) & & 0 \\ & \ddots & \\ 0 & & P(z_n) \end{pmatrix} v$$

hence $\|P(T)\| = \max_{j \leq n} |P(z_j)| \leq \|P\|_\infty$, so that we have (vN) in that case. Now assume that T is merely a contraction on ℓ_2^n. By the polar decomposition, we have $T = U|T|$ with U unitary and $|T|$ hermitian. Note that $\|T\| = \| \, |T| \, \|$. Moreover, $|T|$ is diagonalizable so that there is v unitary such that

$$|T| = v^* \begin{pmatrix} \lambda_1 & & 0 \\ & \ddots & \\ 0 & & \lambda_n \end{pmatrix} v \text{ with } 0 \leq \lambda_j \leq 1. \text{ Let us denote for all } z_1, \ldots, z_n \text{ in}$$

$\overline{D}$

$$T(z_1, \ldots, z_n) = U v^* \begin{pmatrix} z_1 & & 0 \\ & \ddots & \\ 0 & & z_n \end{pmatrix} v$$

so that $T(\lambda_1, \ldots, \lambda_n) = T$.

Let P be any polynomial. Then there are polynomials $p_{ij}(z_1, \ldots, z_n)$ $(i, j = 1, 2, \ldots, n)$ analytic in each variable such that

$$P(T(z_1, \ldots, z_n)) = (p_{ij}(z_1, \ldots, z_n))_{ij}.$$

This shows that the function

$$(z_1, \ldots, z_n) \to \|P(T(z_1, \ldots, z_n))\|$$

is subharmonic in each variable, hence by the maximum principle (applied repeatedly in each variable) we have in particular

$$\|P(T(\lambda_1, \ldots, \lambda_n))\| \le \sup\{\|P(T(z_1, \ldots, z_n))\|, |z_1| = \cdots = |z_n| = 1\}.$$

But by the first part of the argument, since $T(z_1, \ldots, z_n)$ is unitary when $|z_1| = \cdots = |z_n| = 1$, this yields

$$\|P(T)\| \le \|P\|_\infty,$$

hence we obtain (vN) in the matrix case.

We now consider an infinite dimensional Hilbert space H. (I am grateful to Vania Mascioni for correcting a misconception in a previous version.) For simplicity of notation assume H separable and let $\{E_n\}$ be an increasing sequence of finite dimensional subspaces such that $\overline{\cup E_n} = H$. Let P_n be the orthogonal projection from H onto E_n and let $T_n = P_n T P_n$. Clearly $\|T_n\| \le \|T\|$ hence by the first part of the proof for any polynomial P we have

$$\forall n \qquad \|P(T_n)\| \le \|P\|_\infty.$$

Now observe that, for any x in H, $T_n x \to Tx$ when $n \to \infty$. Since $\sup \|T_n\| \le 1$, $T_n \to T$ uniformly over all compact subsets of H, hence $T_n^2 x \to T^2 x$ when $n \to \infty$ and more generally for all k, $T_n^k x \to T^k x$. Consequently, for any polynomial P we have

$$\forall x \in H \qquad P(T_n)x \to P(T)x \quad \text{when} \quad n \to \infty,$$

and we conclude that $\|P(T)\| \le \|P\|_\infty$. $\qquad\qquad\qquad\qquad\qquad\qquad\square$

Usually (vN) is deduced from Sz. Nagy's dilation theorem which is the following.

Theorem 1.1. *Let $T\colon H \to H$ be a contraction. Then there is a Hilbert space $\tilde{H}$ containing H isometrically as a subspace and a unitary operator $U\colon \tilde{H} \to \tilde{H}$ such that*

$$(1.1) \qquad\qquad \forall n \ge 0 \qquad T^n = P_H U^n_{|H}.$$

*When this holds, U is called a **dilation** of T (one also says that U dilates T). The term "strong dilation" is sometimes used in the literature for the same notion.*

Proof: We follow the classical argument of [SNF]. For any n in $\mathbb{Z}$ let $H_n = H$, and consider the Hilbertian direct sum $\tilde{H} = \bigoplus_{n \in \mathbb{Z}} H_n$. On $\tilde{H}$ we introduce the operator $U\colon \tilde{H} \to \tilde{H}$ defined by the following matrix with operator coefficients

$$U = \begin{pmatrix} \ddots & & & & & & \\ & \ddots & I & & & & \\ & & 0 & I & & & \\ & & & 0 & D_T & -T^* & & \\ & & & & T & D_{T^*} & & \\ & & & & 0 & I & & \\ & & & & & 0 & I & \\ & & & & & & \ddots & \ddots \end{pmatrix}$$

where T stands as the $(0,0)$-entry. Equivalently any $(h_n)_{n \in \mathbf{Z}}$ is mapped into $U[(h_n)_{n \in \mathbf{Z}}] = (h'_n)_{n \in \mathbf{Z}}$ with h'_n defined by

(*)
$$h'_n = \begin{cases} h_{n+1} & \text{if} \quad n \notin \{-1, 0\} \\ D_T h_0 - T^* h_1 & \text{if} \quad n = -1 \\ T h_0 + D_{T^*} h_1 & \text{if} \quad n = 0. \end{cases}$$

Let us denote

$$D_T = (1 - T^*T)^{\frac{1}{2}} \quad \text{and} \quad D_{T^*} = (1 - TT^*)^{\frac{1}{2}}.$$

We identify H with $H_0 \subset \tilde{H}$ so that we have $P_H U_{|H} = T$ and more generally $P_H U^n_{|H} = T^n$ for all $n \geq 0$ (note that U has a triangular form, so the diagonal coefficients of U^n are the obvious ones).
We claim that for all $(h_n)_{n \in \mathbf{Z}}$ in $\tilde{H}$ and $(h'_n)_{n \in \mathbf{Z}} = U[(h_n)_{n \in \mathbf{Z}}]$ as above we have

$$\|h'_{-1}\|^2 + \|h'_0\|^2 = \|h_0\|^2 + \|h_1\|^2.$$

Indeed, first note the following identities

$$T^* D_{T^*} = D_T T^* \quad (\text{and} \quad T D_T = D_{T^*} T).$$

(Note that $D_{T^*} = f(TT^*)$ and $D_T = f(T^*T)$ with f continuous and for any polynomial P we have

$$T^* P(TT^*) = P(T^*T)T^*, \quad TP(T^*T) = P(TT^*)T,$$

so approximating f by polynomials we obtain these identities). Then developing $\|h'_{-1}\|^2 + \|h'_0\|^2$ using (*) and the preceding identities, we obtain the above claim. As a consequence, we find that U is an isometry. Moreover U is surjective since it is easy to invert U. Given $h' = (h'_n)_{n \in \mathbf{Z}}$ in $\tilde{H}$, we have $h' = Uh$ with $h = (h_n)_{n \in \mathbf{Z}}$ defined by $h_n = h'_{n-1}$ if $n \notin \{0, 1\}$, $h_0 = D_T h'_{-1} + T^* h'_0$ and $h_1 = -T h'_{-1} + D_{T^*} h'_0$. Equivalently it is clear that U is invertible from the following identity for 2×2 matrices with operator entries

$$\begin{pmatrix} I & 0 \\ 0 & I \end{pmatrix} = \begin{pmatrix} D_T & -T^* \\ T & D_{T^*} \end{pmatrix} \begin{pmatrix} D_T & T^* \\ -T & D_{T^*} \end{pmatrix} = \begin{pmatrix} D_T & T^* \\ -T & D_{T^*} \end{pmatrix} \begin{pmatrix} D_T & -T^* \\ T & D_{T^*} \end{pmatrix}.$$

Therefore we conclude that U is a surjective isometry, hence a unitary operator.

□

From this result, (vN) is easy to derive. Indeed, first observe that if U is unitary we do have for any polynomial P

$$\|P(U)\| \le \|P\|_\infty,$$

since by spectral theory $\|P(U)\| = \sup\{|P(z)| \mid z \in \mathrm{Spec}(U)\}$ and $\mathrm{Spec}(U) \subset \partial D$. Now if T satisfies (1.1) we have $P(T) = P_H P(U)_{|H}$ hence

$$\|P(T)\| \le \|P(U)\| \le \|P\|_\infty$$

and this proves (vN).

Remark. When U is unitary, the spectral functional calculus shows that if $f(z) = P(z, \bar{z})$ is a polynomial in z and $\bar{z}$ we can define $f(U) = P(U, U^*)$ in the obvious way and we still have

$$\|f(U)\| \le \|f\|_\infty = \sup_{|z|=1} |f(z)|.$$

Note that if U is an $n \times n$ matrix, this can be checked by the above argument using the diagonalization of U.

Let us denote by C (resp. $A(D)$) the space of all continuous functions on $\mathbf{T} = \partial D$ (resp. the closed linear span in C of the functions $\{e^{int} \mid n \ge 0\}$). We equip C (or $A(D)$) with the sup norm which we denote (as above) by $\| \ \|_\infty$. Note that $A(D)$ is a subalgebra of C, it is called the disc algebra. Clearly, the functions f of the form $f(z) = P(z, \bar{z})$ for some polynomial P in two variables are dense in C. Therefore by density the linear map

$$f \to f(U)$$

extends to the whole of C. Obviously, we have

(1.2) $$(fg)(U) = f(U)g(U)$$

whenever f and g are of this form, hence this remains true on the completion and we obtain a homomorphism

$$u_U \colon C \to B(H)$$

such that $u_U(\bar{f}) = u_U(f)^*$ and also $\|u_U\| = 1$. This shows that u_U is actually a ∗-representation.

Let us now return to a general contraction T on H. By definition of $A(D)$, the set of all (analytic) polynomials $P(z)$ is dense in $A(D)$. Hence the linear mapping $P \to P(T)$ (which is of norm 1 by (vN)) extends uniquely to a map $u_T \colon A(D) \to B(H)$ such that $\|u_T\| \le 1$. Moreover, again the multiplicativity of u_T is preserved so that we have

(1.3) $$\forall f, g \in A(D) \qquad u_T(fg) = u_T(f)u_T(g).$$

It is customary to write $f(T)$ for $u_T(f)$ whenever f is in $A(D)$, so that we can write

(1.3)' $$\forall f \in A(D) \quad \|f(T)\| \leq \|f\|_\infty.$$

In conclusion, von Neumann's inequality leads to a norm 1 homomorphism $u_T: A(D) \to B(H)$ which maps the function e^{int} to T^n for all $n \geq 0$. It is natural to wonder whether (vN) extends to the case of polynomials in several complex variables.

Quite surprisingly, it turns out that this is possible in 2 variables but impossible in 3 variables. We start by the 2-variable case.

Theorem 1.2. *(Ando's theorem [An1]). Let T_1, T_2 be two commuting contractions on H, i.e. we have*

$$\|T_1\| \leq 1 \quad \|T_2\| \leq 1 \quad \text{and} \quad T_1 T_2 = T_2 T_1.$$

Then H can be embedded isometrically into a Hilbert space $\tilde{H}$ on which there are commuting unitary operators U_1, U_2 such that

$$\forall n, k \geq 0 \qquad P_H U_1^n U_2^k{}_{|H} = T_1^n T_2^k.$$

Consequently, for any polynomial $P(z_1, z_2)$ in two variables we have

$$\|P(T_1, T_2)\| \leq \sup\{|P(z_1, z_2)| \mid z_1, z_2 \in \partial D\}.$$

Remark. Of course the second part follows from the first one exactly as (vN) follows from Sz. Nagy's dilation theorem. Just observe

$$U_1 U_2 = U_2 U_1 \Rightarrow U_2 U_1^{-1} = U_1^{-1} U_2$$

and $U_1^{-1} = U_1^*$, hence not only U_1, U_2 but also U_1, U_2, U_1^*, U_2^* all mutually commute hence lie in a commutative unital C^*-subalgebra $A \subset B(H)$. By Gelfand's theorem A is isomorphic to $C(K)$, where K is the set of all homomorphisms

$$\chi: A \to \mathbb{C} \quad \text{such that} \quad \chi(1) = \|\chi\|.$$

We have then
$$\|P(U_1, U_2)\| = \sup_{\chi \in K} |\langle \chi, P(U_1, U_2)\rangle|$$
$$= \sup_{\chi \in K} |P(\chi(U_1), \chi(U_2))|$$

hence since $|\chi(U_1)| = |\chi(U_2)| = 1$

$$\|P(U_1, U_2)\| \leq \sup_{\substack{|z_1|=1 \\ |z_2|=1}} |P(z_1, z_2)|.$$

Before proving Ando's theorem, we first gather more information on isometries. An operator $T: H \to H$ is called an isometry if $\|Tx\| = \|x\|$ for all x in H.

Equivalently, this means that $T^*T = I$. In general an isometry is not surjective. An isometry is surjective iff it is a unitary operator. The fundamental example of an isometry is the shift operator on the space $\ell_2(\mathbb{N}, H) = H \oplus H \oplus \cdots$ (direct sum of a collection of copies of H indexed by $\mathbb{N}$), namely the operator s on $\ell_2(\mathbb{N}, H)$ which maps $(x_0, x_1, x_2, \ldots)$ to $(0, x_0, x_1, x_2, \ldots)$. We will call a shift any operator of this kind for some Hilbert space H. (The dimension of H is called the multiplicity of the shift.) The next result is quite important. It shows that any isometry can be decomposed as the direct sum of a shift as above and a unitary operator, this is called the Wold decomposition.

Theorem 1.3. Let $V \colon H \to H$ be an isometry. Let $H_\infty = \bigcap\limits_{n \geq 0} V^n(H)$ and let $H_0 = H \ominus H_\infty$ so that $H = H_0 \oplus H_\infty$. Then H_∞ and H_0 are reducing subspaces for V (i.e. they are invariant under V and V^*) such that

(i) $V_{|H_\infty} \colon H_\infty \to H_\infty$ is a unitary operator from H_∞ onto itself.
(ii) $V_{|H_0} \colon H_0 \to H_0$ is unitarily equivalent to a shift as above with multiplicity equal   to $\dim(H \ominus V(H))$.

Proof: The inclusion $V(H_\infty) \subset H_\infty$ is clear. Moreover if $x = V^{n+1}y$ then $V^*x = V^*VV^ny = V^ny$ hence $V^*(H_\infty) \subset H_\infty$. This implies that H_∞ and H_0 are reducing subspaces for V. Moreover, clearly $V(H_\infty) = H_\infty$, hence $V_{|H_\infty} \colon H_\infty \to H_\infty$ is a surjective isometry, hence is unitary. This proves (i). To check (ii) note that

$$H_\infty^\perp = \overline{\operatorname{span}}\left(\bigcup_n V^n(H)^\perp\right) = \overline{\operatorname{span}}\left(\bigcup_n E_n\right)$$

where $E_n = V^n(H) \ominus V^{n+1}(H)$ for all $n \geq 0$. Note that since V is an isometry it preserves angles hence we have $E_n = V^n(E_0)$ for all $n \geq 0$ and

$$V^n(E_0) \perp V^m(E_0) \quad \text{for all} \quad n \neq m.$$

It follows that $H_\infty^\perp$ is isometrically isomorphic to the space $\ell_2(E_0)$. Indeed, the map $U \colon \ell_2(E_0) \to H_\infty^\perp$ defined by $U(x_0, x_1, \ldots) = \sum\limits_{n \geq 0} V^n x_n$ is a unitary operator. Finally we clearly have (with the preceding notation) $V = UsU^{-1}$. This completes the proof of (ii).   □

Let H be a Hilbert space and let $v \colon E \to E$ be an operator defined on a subspace $E \subset H$. We will say that an operator $V \colon H \to H$ extends v if $V(E) \subset E$ and $V(x) = v(x)$ for all x in E.

As an immediate consequence of Theorem 1.3 we have

Lemma 1.4. Any isometry $V \colon H \to H$ on a Hilbert space can be extended to a unitary operator, i.e. there is a Hilbert space $\tilde{H}$ containing H isometrically and a unitary operator $U \colon \tilde{H} \to \tilde{H}$ which extends V.

Proof: If V is a shift s on $\ell_2(\mathbb{N}, H)$, we can take for U the so-called "bilateral shift" on $\ell_2(\mathbb{Z}, H)$ which is unitary and clearly extends V. On the other hand if V

is unitary the above statement is trivial. By the decomposition in Theorem 1.3, the general case reduces to these two cases.   □

Remark. If we wish we can always ensure that the unitary operator U is minimal i.e. such that the subspace $E = \text{span}\left(\bigcup_{n \in \mathbb{Z}} U^n(H)\right)$ is dense in $\tilde{H}$. Indeed, if E is not dense, we simply replace $\tilde{H}$ by $\overline{E}$ and note that E is U-invariant and $U_{|\overline{E}}$ is unitary.

Lemma 1.5. *Let V_1, V_2 be two commuting isometries on a Hilbert space H, then there is a Hilbert space $\tilde{H}$ and commuting unitary operators U_1, U_2 on $\tilde{H}$ admitting H as an invariant subspace and such that*

$$U_{1|H} = V_1 \quad \text{and} \quad U_{2|H} = V_2.$$

In other words, two commuting isometries can be extended to two commuting unitaries.

Remark. Actually it is easy to modify the argument to yield (by an inductive reasoning) that the same is true for any number (and even any set) of commuting isometries.

Proof of Lemma 1.5. Let $H \subset \tilde{H}$ and let $U_1 \colon \tilde{H} \to \tilde{H}$ be a *minimal* unitary extension of V_1 obtained by applying Lemma 1.4 and the remark following it, i.e. we assume that the subspace E formed of all the finite sums of the form

$$\sum_{n \in \mathbb{Z}} U_1^n h_n, \qquad h_n \in H$$

is dense in $\tilde{H}$.

This will allow us to define an extension $\tilde{V}_2$ of V_2 which commutes with U_1 and we will do this without "spoiling" the good properties of V_2. If we want $\tilde{V}_2$ to commute with U_1 and to extend V_2, we *must* define $\tilde{V}_2$ on E as follows

$$\tilde{V}_2\left(\sum_{n \in \mathbb{Z}} U_1^n h_n\right) = \sum_{n \in \mathbb{Z}} U_1^n (V_2 h_n).$$

To verify that this definition is unambiguous and at the same time that it defines an isometry, we observe that

$$\left\|\sum U_1^n V_2 h_n\right\|^2 = \sum_{n,m} \langle U_1^n V_2 h_n, U_1^m V_2 h_m\rangle$$

$$= \sum_{n \geq m} \langle U_1^{n-m} V_2 h_n, V_2 h_m\rangle + \sum_{n < m} \langle V_2 h_n, U_1^{m-n} V_2 h_m\rangle$$

hence since U_1 extends V_1

$$= \sum_{n \geq m} \langle V_1^{n-m} V_2 h_n, V_2 h_m\rangle + \sum_{n < m} \langle V_2 h_n, V_1^{m-n} V_2 h_m\rangle,$$

since V_2 and V_1 commute

$$= \sum_{n \geq m} \langle V_2 V_1^{n-m} h_n, V_2 h_m \rangle + \sum_{n < m} \langle V_2 h_n, V_2 V_1^{m-n} h_m \rangle$$

since V_2 is an isometry

$$= \sum_{n \geq m} \langle V_1^{n-m} h_n, h_m \rangle + \sum_{n < m} \langle h_n, V_1^{m-n} h_m \rangle$$

and finally (applying the preceding calculation for $V_2 = I$)

$$= \left\| \sum U_1^n h_n \right\|^2 .$$

This shows that $\tilde{V}_2$ is an isometry and is unambiguously defined. Moreover, $\tilde{V}_2 h_0 = V_2 h_0$ for all h_0 in H hence $\tilde{V}_2(H) \subset H$ and $\tilde{V}_2$ extends V_2. Finally, the important point is the following: we claim that if V_2 already is unitary, then $\tilde{V}_2$ still is unitary. Indeed, recall that an isometry is unitary iff it is surjective, or equivalently iff it has a dense range. But if $V_2 \colon H \to H$ is surjective, clearly the operator $\tilde{V}_2$ that we just defined has dense range (its range contains E), hence is unitary. This proves our claim. Hence if V_2 happens to be unitary, $\tilde{V}_2$ also is and we are done. To complete the proof in case V_2 is not unitary, we simply repeat the construction applying Lemma 1.4 this time to the isometry $\tilde{V}_2$ instead of V_1. By the above claim we can get a unitary extension of $\tilde{V}_2$ and a (still) unitary extension of U_1. This gives the desired result.   □

The following proof is essentially the original one from [An1].

Proof of Theorem 1.2. By Lemma 1.5, it suffices to show that two commuting contractions can be dilated to two commuting isometries (this is the step which is restricted to "two"!). Let T_1, T_2 be mutually commuting contractions on a Hilbert space H. Let $H_+ = H \oplus H \oplus \cdots$ (or $H = \ell_2(\mathbb{N}, H)$) be the direct sum of a family of copies of H indexed by $\mathbb{N}$. Let $W_1 \colon H_+ \to H_+$ and $W_2 \colon H_+ \to H_+$ be the operators defined by

$$\forall\, h = (h_0, h_1, \ldots) \qquad W_i h = (T_i h_0, D_{T_i} h_0, 0, h_1, h_2, \ldots)$$

where $i = 1, 2$ and where $D_{T_i} = (1 - T_i^* T_i)^{1/2}$ as above in the proof of Sz. Nagy's dilation theorem. Clearly W_1 and W_2 are isometries on H_+, but in general they do not commute. We will modify them to obtain commuting isometries. Let $H_4 = H \oplus H \oplus H \oplus H$ and let us identify H_+ and $H \oplus H_4 \oplus H_4 \oplus H_4 \oplus \cdots$ by the following identification:

$$h = (h_n)_{n \geq 0} \longrightarrow (h_0, \{h_1, \ldots, h_4\}, \{h_5, \ldots, h_8\}, \ldots).$$

On the space H_4, consider a unitary operator v (to be specified later) and let $V \colon H_+ \to H_+$ be defined for all $h = (h_0, k_1, k_2, \ldots)$, $h_0 \in H$, $k_1 \in H_4$, $k_2 \in H_4, \ldots$ by the following formula

$$V h = (h_0, v k_1, v k_2, \ldots).$$

Clearly V is unitary and

$$V^{-1} h = (h_0, v^{-1} k_1, v^{-1} k_2, \ldots).$$

We define $V_1 = VW_1$ and $V_2 = W_2 V^{-1}$. Clearly these are isometries on H_+. The key point is that v can be chosen in order to ensure that V_1 and V_2 commute. To check this let us compute $V_1 V_2 h$ and $V_2 V_1 h$ for an element $h = (h_0, k_1, k_2, \ldots)$ in H_+ with $h_0 \in H$, $k_1 \in H_4$, $k_2 \in H_4$, etc. By a simple calculation we find

$$V_1 V_2 h = (T_1 T_2 h_0, v\{D_{T_1} T_2 h_0, 0, D_{T_2} h_0, 0\}, k_1, k_2, \ldots)$$

and

$$V_2 V_1 h = (T_2 T_1 h_0, \{D_{T_2} T_1 h_0, 0, D_{T_1} h_0, 0\}, k_1, k_2, \ldots).$$

Hence (since $T_1 T_2 = T_2 T_1$) if we want $V_1 V_2 = V_2 V_1$ we *must* define

$$\forall\, h_0 \in H \qquad v(\{D_{T_1} T_2 h_0, 0, D_{T_2} h_0, 0\}) = \{D_{T_2} T_1 h_0, 0, D_{T_1} h_0, 0\}.$$

By a simple computation, the reader will check that

$$\{D_{T_2} T_1 h_0, 0, D_{T_1} h_0, 0\} \quad \text{and} \quad \{D_{T_1} T_2 h_0, 0, D_{T_2} h_0, 0\}$$

have the same norm in H_4, hence this defines v as an isometry from the subspace

$$\Lambda_1 = \{(D_{T_1} T_2 h_0, 0, D_{T_2} h_0, 0) \mid h_0 \in H\}$$

onto the subspace

$$\Lambda_2 = \{(D_{T_2} T_1 h_0, 0, D_{T_1} h_0, 0) \mid h_0 \in H\}.$$

By density, of course v defines an isometry from the closure of Λ_1 onto the closure of Λ_2. To extend v to an isometry of H_4 *onto* itself (i.e. to a unitary operator on H_4) it suffices to check that $H_4 \ominus \Lambda_1$ and $H_4 \ominus \Lambda_2$ have the same (Hilbertian) dimension. If H_4 is finite dimensional, this is clear since Λ_1 and Λ_2 are isometric, and if H_4 is infinite dimensional it is equally clear since the dimension of $H_4 \ominus \Lambda_1$ and $H_4 \ominus \Lambda_2$ are at most $\dim H_4 = \dim H$ and the zero coordinates in the definition of Λ_1 and Λ_2 ensures that they are at least $\dim H$. This shows that v can be chosen so that V_1 and V_2 commute. Thus we obtain commuting isometries V_1, V_2 which dilate T_1, T_2. Applying Lemma 1.5 to V_1, V_2 we obtain the conclusion. $\square$

Remark. Here is a third route to prove (vN). (This was communicated to me by J. Arazy, this proof appears already in Drury's survey [Dr]). This approach is based on the fact [Fi, R2] that the unit ball of the disc algebra $A(D)$ is the closed convex hull of the set of finite Blaschke products. Taking this for granted for the moment, it is easy to see that (vN) reduces to proving $\|f(T)\| \leq 1$ when f is a finite Blaschke product, or merely when φ is a Blaschke "factor" i.e. a Möbius map of the form

$$\varphi_\lambda(z) = \frac{z - \lambda}{1 - \bar{\lambda} z} \quad \text{for} \quad |\lambda| < 1.$$

But for such maps (vN) is easy: indeed by a simple calculation we find if $T \in B(H)$, $\|T\| \leq 1$ and $x \in H$

$$\|(T - \lambda)x\|^2 - \|(I - \bar{\lambda} T)x\|^2 = (\|Tx\|^2 - \|x\|^2)(1 - |\lambda|^2) \leq 0$$

hence $\|(T - \lambda)(1 - \bar{\lambda}T)^{-1}\| \leq 1$ which means that $\|\varphi_\lambda(T)\| \leq 1$. This completes our third proof of (vN). (Note that we need to know that $f \to f(T)$ is a priori continuous to pass to the *closure* of the convex hull, but this can be a priori guaranteed by replacing T by rT with $r < 1$, and by letting $r \to 1$ in the end.)

Now let us verify that the unit ball of $A(D)$ is the closed convex hull of the finite Blaschke products. Fortunately, there is a very elegant and simple proof of this due to Alain Bernard (cf. [BGM] for extensions), as follows. It clearly suffices to show that any polynomial f in $A(D)$ with $\|f\| < 1$ lies in the closed convex hull of the finite Blaschke products. Let f be such a polynomial. Choosing $g(z) = z^N$ with $N = \deg(f)$ we find an analytic polynomial g of modulus one on ∂D such that $g\bar{f} \in A(D)$.

Consider then, for any real number t, the function

$$f_t = \frac{f + e^{it}g}{1 + e^{it}g\bar{f}}.$$

Note that f_t is clearly analytic. We claim that f_t is of modulus one on ∂D. Indeed, if φ_λ denotes the Möbius map as above then we can write

$$f_t(z) = \varphi_{-f(z)}(e^{it}g(z))$$

which implies that $|f_t(z)| = 1$ when $|z| = 1$. In other words, f_t is inner. Since it is continuous on $\overline{D}$, it must be a finite Blaschke product for each real t. (Note that f_t has at most N zeros.)

Finally, viewed as a function of e^{it}, f_t is the boundary value of the analytic function

$$\xi \to \frac{f + \xi g}{1 + \xi g\bar{f}}$$

(defined for $\xi \in D$), hence we have (taking $\xi = 0$)

$$f = \int f_t \frac{dt}{2\pi}.$$

Then, a suitable discretization of this average shows that f lies in the closed convex hull of the finite Blaschke products.   □

Remark. There is a very striking analogy between the preceding argument and the known proofs of the Russo-Dye Theorem in the C^*-algebra case (see e.g. the proof in [Ped, p.4]). Concerning the latter proof, it is important to observe that, when we represent a point in the open unit ball of say $B(H)$ as the barycenter of a probability measure supported on the unitaries, the "representing" measure is actually a Jensen measure (as in the preceding remark), so that the barycenter formula is valid not only for affine functions but also for analytic ones. This leads to one more transparent proof for von Neumann's inequality.

The problem to extend Ando's theorem to three (or more) mutually commuting contractions remained open for a while until Varopoulos [V1] found a counterexample (another example was given by Crabb-Davie [CD]). Of course this implies that Ando's dilation theorem (the first part of the preceding Theorem 1.2) is not valid for three mutually commuting contractions. For an explicit construction of three mutually commuting contractions (T_1, T_2, T_3) which do not dilate to three commuting unitaries, see [Par1]. Moreover, Parrott's example satisfies

$$\|P(T_1, T_2, T_3)\| \leq \sup\{|P(z_1, z_2, z_3)| \mid (z_1, z_2, z_3) \in D^3\}.$$

In other words, even though it does not dilate, the triple (T_1, T_2, T_3) does satisfy the von Neumann inequality. This shows that Parrott's example is quite different from the ones of Varopoulos or Crabb-Davie. See the notes on Chapter 4, for variants of Parrott's example.

We will describe one of Varapoulos's counter examples in chapter 5. The Crabb-Davie example in [CD] is very simple. The three commuting contractions act on an 8-dimensional space and the polynomial is homogeneous of degree 3. Let us briefly describe this example. Let H be an 8-dimensional Hilbert space with orthonormal basis $e, f_1, f_2, f_3, g_1, g_2, g_3, h$. We define the operators on H by letting (for $i, j, k = 1, 2, 3$) $T_i e = f_i$, $T_i f_i = -g_i$, $T_i f_j = g_k$ (if $i \neq j$, with $k \neq i$ and $k \neq j$), $T_i g_j = \delta_{ij} h$ and $T_i h = 0$. We observe that

(1.4)
$$T_i T_j f_k = \begin{cases} -h & \text{if } i = j = k \\ h & \text{if } i, j, k \text{ are all different} \\ 0 & \text{otherwise.} \end{cases}$$

Then it is easy to check that T_1, T_2, T_3 are mutually commuting contractions. Let P be the homogeneous polynomial of degree 3 defined by

$$P(z_1, z_2, z_3) = z_1 z_2 z_3 - z_1^3 - z_2^3 - z_3^3.$$

It is any easy exercise to check that $\|P\|_\infty < 4$, but on the other hand we have $T_i T_j T_k e = T_i T_j f_k$, hence (1.4) yields $P(T_1, T_2, T_3)e = 4h$ therefore $\|P(T_1, T_2, T_3)\| \geq 4 > \|P\|_\infty$. This yields the announced example.

However, rather surprisingly the following is apparently still open. Let $n \geq 1$. Let us introduce the (possibly infinite) constant

$$C_n = \sup\{\|P(T_1, \ldots, T_n)\|\}$$

where the supremum runs over all mutually commuting contractions $T_1, \ldots, T_n$ on H (infinite dimensional Hilbert space) and over all polynomials P such that $|P(z_1, \ldots, z_n)| \leq 1$ for all $(z_1, \ldots, z_n)$ in $(\partial D)^n$. We have just seen that $C_1 = C_2 = 1$ and we mentioned (cf. [V1, CD]) that $C_3 > 1$.

Problem. Is C_3 finite? Is C_n finite for any $n \geq 3$? (Note that C_n is clearly nondecreasing.) The following observation is certainly well known.

Proposition 1.6. *For all $n, m \geq 1$, $C_{n+m} \geq C_n C_m$. Therefore $C_n \to \infty$ when $n \to \infty$.*

Proof: We may assume that C_n and C_m are finite otherwise C_{n+m} is clearly also infinite. Fix $\varepsilon > 0$. Let $T_1, \ldots, T_n$ and P be as in the definition of C_n with

$$\|P(T_1,\ldots,T_n)\| > C_n - \varepsilon.$$

Similarly consider commuting contractions $T_{n+1},\ldots,T_{n+m}$ and a polynomial $Q(z_{n+1},\ldots,z_{n+m})$ with $|Q| \le 1$ on $(\partial D)^m$ such that $\|Q(T_{n+1},\ldots,T_{n+m})\| > C_m - \varepsilon$. Let $\widehat{T}_j = T_j \otimes I_H$ for $j = 1,2,\ldots,n$ and $\widehat{T}_j = I_H \otimes T_j$ for $j = n+1,\ldots,n+m$. Then $\{\widehat{T}_j \mid j = 1,\ldots,n+m\}$ are $(n+m)$ contractions which commute and if we let

$$R(z_1,\ldots,z_{n+m}) = P(z_1,\ldots,z_n)Q(z_{n+1},\ldots,z_{n+m})$$

we find

$$R(\widehat{T}_1,\ldots,\widehat{T}_{n+m}) = P(T_1,\ldots,T_n) \otimes Q(T_{n+1},\ldots,T_{n+m}),$$

hence

$$\|R(\widehat{T}_1,\ldots,\widehat{T}_{n+m})\| = \|P(T_1,\ldots,T_n)\|\cdot\|Q(T_{n+1},\ldots,T_{n+m})\| \ge (C_n-\varepsilon)(C_m-\varepsilon).$$

Thus we conclude $C_{n+m} \ge (C_n-\varepsilon)(C_m-\varepsilon)$. In particular, $C_{3n} \ge C_3{}^n \to \infty$. $\quad\square$

Let $U\colon H \to H$ be a unitary operator. It is natural to wonder what kind of subspaces $E \subset H$ have the property that if $T = P_E U_{|E}$ we have

$$\forall n \ge 2 \qquad T^n = P_E U^n_{|E}.$$

In other words, when is U a (strong) dilation of $T = P_E U_{|E}$? The operator $P_E U_{|E}$ is usually called the compression of U to E. Actually this question makes sense even if U is not unitary. More generally, we may consider a bounded homomorphism $\pi\colon A \to B(H)$ defined on a Banach algebra A and ask for which subspaces $E \subset H$ the map

(1.5) $$\pi_E\colon a \to P_E \pi(a)_{|E}$$

is a homomorphism. One obvious example of this situation is when we have

$$\forall\, a \in A \qquad \pi(a)E \subset E,$$

and in that case we say that E is π invariant.

Furthermore if E_2, E_1 are π-invariant (closed) subspaces with $E_2 \subset E_1$, it can be checked that for $E = E_1 \ominus E_2$ the map π_E is a homomorphism (although in general E is not π-invariant). Such spaces of the form $E = E_1 \ominus E_2$ are usually called semi-invariant, we will call them π-semi-invariant when we want to specify which homomorphism is involved. When considering a single operator T in $B(H)$ (or the algebra it generates) we will say that a subspace E is semi-invariant for T if there are invariant subspaces $E_2 \subset E_1$ (i.e. subspaces such that $T(E_j) \subset E_j$, $j = 1,2$) such that $E = E_1 \ominus E_2$. The following striking discovery of Sarason answers the above questions in a very broad context. (It could be stated as a purely algebraic fact.)

Theorem 1.7. *Let A be a Banach algebra and let $\pi\colon A \to B(H)$ be a bounded homomorphism. Let $E \subset H$ be a (closed) subspace of H. For a in A, let $\pi_E(a) = P_E\pi(a)_{|E}$. Note $\pi_E(a) \in B(E)$. Then π_E is a homomorphism from A into $B(E)$ iff there are π-invariant closed subspaces $E_2 \subset E_1$ such that*

$$E = E_1 \ominus E_2.$$

Proof: Consider first $E = E_1 \ominus E_2$. Observe that $E_1 \ominus E_2$ can be identified with E_1/E_2. Let $S\colon E_1/E_2 \to E_1 \ominus E_2$ be the canonical isometry. Clearly the map $\pi_j\colon A \to B(E_j)$ defined for $j = 1,2$ by $\pi_j(a) = \pi(a)_{|E_j}$, is a homomorphism. Let $Q\colon E_1 \to E_1/E_2$ be the quotient map and consider the map $Q\pi_1(a)\colon E_1 \to E_1/E_2$. Since $\mathrm{Ker}(Q\pi_1(a)) \supset E_2$, there is a uniquely determined map $\tilde{\pi}(a)\colon E_1/E_2 \to E_1/E_2$ such that $\tilde{\pi}(a)Q = Q\pi_1(a)$, and the unicity and the fact that π_1 is a homomorphism clearly imply that $\tilde{\pi}(a)\tilde{\pi}(b) = \tilde{\pi}(ab)$ for all a, b in A. Hence $\tilde{\pi}$ is a homomorphism. Now it is easy to check that

$$\pi_E(a) = S\tilde{\pi}(a)S^{-1},$$

hence π_E itself is a homomorphism.

Conversely, let $E \subset H$ be a subspace such that π_E defined by (1.5) is a homomorphism.

Let E_1 be the closed linear span of $E \cup \left(\bigcup_{a \in A} \pi(a)E \right)$. Then E_1 is obviously a (closed) π-invariant subspace and $E_1 \supset E$. Let $E_2 = E_1 \ominus E$. To conclude, it suffices to check that E_2 also is π-invariant. For that purpose, let $x \in E_1 \ominus E = E_2$ and let $a \in A$. We will prove that $\pi(a)x \in E_2$. For each n, there are finite sets (a_i^n) in A, (x_i^n) in E and x^n in E such that

$$x = \lim_{n \to \infty} \left(x^n + \sum_i \pi(a_i^n)x_i^n \right).$$

Note that $P_E x = 0$ since $x \in E_1 \ominus E$, so that

$$(1.6) \qquad 0 = P_E x = \lim_{n \to \infty} \left(x^n + \sum_i \pi_E(a_i^n)x_i^n \right).$$

To check that $\pi(a)x \in E_2$, it suffices to show that $\pi(a)x \in E^\perp$. We have then

$$P_E\pi(a)x = \lim_{n \to \infty} \left(P_E\pi(a)x^n + \sum_i P_E\pi(a)\pi(a_i^n)x_i^n \right)$$

$$= \lim_{n \to \infty} \left(\pi_E(a)x^n + \sum_i \pi_E(aa_i^n)x_i^n \right)$$

hence since π_E is assumed a homomorphism

$$= \lim_{n \to \infty} \left(\pi_E(a)x^n + \sum_i \pi_E(a)\pi_E(a_i^n)x_i^n \right)$$

$$= \lim_{n \to \infty} \pi_E(a) \left(x^n + \sum_i \pi_E(a_i^n)x_i^n \right)$$

hence by (1.6) $P_E\pi(a)x = 0$. This shows that $\pi(a)x \in E^\perp$ and concludes the proof.   □

Remark. In particular this applies to Sz.-Nagy's dilation theorem (or Ando's). For any contraction T there is a unitary operator U admitting invariant subspaces E_2, E_1 with $E_2 \subset E_1$ such that

$$T = P_{E_1 \ominus E_2} U_{|E_1 \ominus E_2}.$$

This explains why the structure of the invariant subspaces of unitary operators is so important to understand the structure of general contractions. See [SNF] for much more on this theme.

Remark 1.8. Sarason's result has a purely algebraic version as follows. Let A be an algebra and let V be a vector space. Consider a homomorphism $\pi\colon A \to L(V)$ (where $L(V)$ is the space of all linear maps on V). Let W be another vector space, and let $w\colon W \to V$ and $v\colon V \to W$ be linear maps such that the map $\tilde{\pi}\colon A \to L(W)$ defined by

$$\forall\, a \in A \qquad \tilde{\pi}(a) = v\pi(a)w$$

is a homomorphism. Assume also that $vw = I$. Then there are π-invariant subspaces $E_2 \subset E_1 \subset V$ and an isomorphism $S\colon E_1/E_2 \to W$ such that the homomorphism

$$a \to S^{-1}\tilde{\pi}(a)S \in L(E_1/E_2)$$

is nothing but the canonical "compression" of π to E_1/E_2. We will come back to this later in the Banach space framework (see Chapter 4).

It is natural to ask whether von Neumann's inequality can be valid on other spaces than Hilbert spaces. The answer is negative, as shown by Foias [Foi].

Theorem 1.9. *Let X be a complex Banach space. Assume that for all $T\colon X \to X$ with $\|T\| = 1$ we have $\|P(T)\| \leq \|P\|_\infty$ for all polynomials P. Then X is isometric to a Hilbert space.*

Proof: Consider λ in D, let

$$\varphi_\lambda(z) = \frac{z - \lambda}{1 - \bar{\lambda}z},$$

be the associated Möbius transformation. Clearly φ_λ is analytic in a neighborhood of $\overline{D}$ and φ_λ preserves ∂D so that $\varphi_{\lambda|\partial D}$ is in A and $\|\varphi_\lambda\|_\infty = 1$. Clearly by the obvious extension of (1.3) if T is a contraction on X we have $\varphi_\lambda(T) = (T - \lambda)(1 - \bar{\lambda}T)^{-1}$. Consider now x, y in X with $\|x\| = \|y\| = 1$. Let x^* be such that $\|x^*\| = \langle x^*, x \rangle = 1$, and let $T = x^* \otimes y$, i.e. T is defined by $T(a) = x^*(a)y$

$\forall a \in X$. Note that $T(x) = y$. Clearly $\|T\| = \|x^*\| \|y\| = 1$ and our assumption implies

$$\|\varphi_\lambda(T)\| = \|(T - \lambda)(1 - \bar{\lambda}T)^{-1}\| \leq 1$$

hence

$$\|(T - \lambda)x\| \leq \|(1 - \bar{\lambda}T)x\|,$$

or

$$\|y - \lambda x\| \leq \|x - \bar{\lambda}y\|.$$

Exchanging x and y as well as λ and $\bar{\lambda}$ we obtain the converse inequality so that

$$\forall \lambda \in D \qquad \|y - \lambda x\| = \|x - \bar{\lambda}y\|.$$

Consequently we have if $t \in [-1, 1]$

$$\|y + tx\| = \|x + ty\|$$

hence if $t \neq 0$

$$\left\|\frac{1}{t}y + x\right\| = \left\|\frac{1}{t}x + y\right\|$$

so that if t is real with $|t| \geq 1$ we also have

$$\|ty + x\| = \|tx + y\|.$$

Finally, we obtain that if $\|x\| = \|y\| = 1$

$$\forall t \in \mathbb{R} \qquad \|x + ty\| = \|tx + y\|.$$

This means that the underlying real Banach space has the property that for any two dimensional subspace E and any two points x, y of the unit sphere of E, there is an isometry of E which maps x to y. It is well known (cf. [Fic] for the original argument, see also [A] for more information) that this implies that X is isometric to a Hilbert space.

Notes and Remarks on Chapter 1

The very simple proof of von Neumann's inequality which opens this chapter is due to Ed Nelson [Nel] (I thought it was new until Vern Paulsen directed me to John Wermer who gave me this reference).

For the rest of the chapter, most references are given in the text. We refer the reader to the classical reference [SNF] for more information.

After the appearance of [SNF], dilation theory became a very important branch of operator theory, see [Pa1] for more recent work in this direction. It is also closely connected with Arveson's theory of nest algebras [see Ar2 and Dad].

After the papers [V1, CD], numerous attempts were made to prove versions of von Neumann's inequality for n-tuples of operators $T_1, \ldots, T_n$ see for instance [Ble, Dix1, DD, Dr, To1-2, Ho4-5-6, Gu]. Note for instance that Drury [Dr, p. 21] observed that the inequality holds for any number of commuting contractions on a 2-dimensional Hilbert space. For recent results relating von Neumann's inequality to the theory of analytic functions in several variables, see [CW, CLW, Na, Lo, LoS].

If we do not assume that the contractions $T_1, \ldots, T_n$ are commuting, then the case of a "free" n-tuple of operators appears as a typical example.

For instance, Bożejko [B4] proved the following extension of von Neumann's inequality: let $P(X_1, \ldots, X_n)$ be an arbitrary polynomial in the non-commuting (formal) variables $X_1, \ldots, X_n$, then for any n-tuples of Hilbert space contractions $(T_1, \ldots, T_n)$ we have

$$\|P(T_1, \ldots, T_n)\| \leq \sup\{\|P(U_1, \ldots, U_n)\|\}$$

where the supremum runs over all possible n-tuples $(U_1, \ldots, U_n)$ of unitary operators on Hilbert space. (Actually, this supremum is already achieved on the set of n-tuples of unitary matrices of arbitrary size.) Although this is not the original proof, this result can be proved using the same idea as in the proof of (vN) given at the beginning of Chapter 1.

In another direction, the paper [Po1] contains an interesting extension of (vN) for n-tuples of operators $T_1, \ldots, T_n$ such that $\left\|\sum_1^n T_i T_i^*\right\| \leq 1$. This can be stated as follows: let $\mathcal{F} = \mathbb{C}1 \oplus \left(\bigoplus_{m \geq 1} H^{\otimes m} \right)$ be the full Fock space with $H = \ell_2^n$. Let $e_1, \ldots, e_n$ be the canonical basis of ℓ_2^n and let $S_j \colon \mathcal{F} \to \mathcal{F}$ be the operator (of "creation of particle") defined by $S_j h = e_j \otimes h$. Then the main result of [Po1] states that for any polynomial P as above

$$\|P(T_1, \ldots, T_n)\| \le \|P(S_1, \ldots, S_n)\|.$$

See also [Po2-4].

There also has been attempts to extend von Neumann's inequality to contractions acting on L_p-spaces. In these extensions, the shift operator

$$S : \ell_p \to \ell_p$$

defined by

$$S(x_0, x_1, x_2, \ldots) = (x_1, x_2, \ldots),$$

plays a crucial rôle. The sup norm of the polynomial $P = \sum a_n z^n$ on the right hand side of (vN) must be replaced by the norm of $P(S)$ acting on ℓ_p. Thus, a famous conjecture due to Matsaev (1966) asserts that for any contraction T on L_p ($1 < p < \infty$) we have, for any polynomial P as above,

$$\|P(T)\|_{L_p \to L_p} \le \|P(S)\|_{\ell_p \to \ell_p}.$$

If $p = 2$, this is von Neumann's inequality, but for $p \ne 2$ it is still open. However, using the dilation theory for *positive* contractions on L_p-spaces (see [AS]), this was proved in [Pe3] and [CRW] with additional assumptions on T. For instance, it suffices that T admits a contractive majorant, *i.e.* there is a positive contraction $\tilde{T}$ on L_p such that $|Tf| \le \tilde{T}(|f|)$ a.e. for all f in L_p, see [Pe2] for a survey. Note however, that Matsaev's conjecture remains wide open: it is not even known if T is a 2 by 2 matrix!

2. Non-unitarizable uniformly bounded group representations

Summary: In this chapter, we introduce the space $B(G)$ of coefficients of unitary representations on a discrete group G and a related space $T_p(G)$, $(1 \leq p < \infty)$ of complex valued functions on G. We show that if $G = \mathbb{F}_N$, the free group on $N \geq 2$ generators, there are non-unitarizable uniformly bounded representations on G. We give several related characterizations of amenable groups. Then we extend the method to the case of semi-groups. This allows us to produce (this time for $G = \mathbb{N}$) examples of power bounded operators which are not polynomially bounded.

This chapter is mainly based on the papers [Fe1, BF2, B2 and W1]. Our aim is to exhibit rapidly an example of a uniformly bounded representation on a group G which is not unitarizable. By Theorem 0.6, we know that the group G cannot be amenable. We will therefore discuss non-amenable groups and in particular the free group with N generators which we will denote by $\mathbb{F}_N$.

Let G be a discrete group. We denote by $\lambda\colon G \to B(\ell_2(G))$ the left regular representation defined by

$$\forall h \in \ell_2(G) \qquad \lambda(t)h = \delta_t * h$$

or equivalently

$$\forall t, s \in G \qquad (\lambda(t)h)(s) = h(t^{-1}s).$$

Similarly we denote by $\rho\colon G \to B(\ell_2(G))$ the right regular representation defined by

$$\forall h \in \ell_2(G) \quad \forall t, s \in G \qquad (\rho(t)h)(s) = (h * \delta_{t^{-1}})(s) = h(st).$$

Note that $\lambda(t)$ (resp. $\rho(t^{-1})$) is the unitary operator of left (resp. right) translation by t on $\ell_2(G)$. Clearly these are unitary representations of G.

Observe that λ and ρ commute, i.e.

(2.1) $$\forall t, s \in G \qquad \lambda(t)\rho(s) = \rho(s)\lambda(t).$$

We will denote by $B(G)$ the space consisting of all the matrix coefficients of the unitary representations of G. More precisely $B(G)$ is the space of all functions $f\colon G \to \mathbb{C}$ for which there is a Hilbert space H, a unitary representation $\pi\colon G \to B(H)$ and elements x, y in H such that

(2.2) $$f(t) = \langle \pi(t)x, y \rangle.$$

It can be checked (exercise) that this is a linear space and that the norm defined by

$$\|f\|_{B(G)} = \inf\{\|x\|\,\|y\|\}$$

is indeed a norm with which $B(G)$ is a Banach space. (Hint: Use the identity

$$\langle \pi_1(t)x_1, y_1 \rangle + \langle \pi_2(t)x_2, y_2 \rangle = \langle \pi(t)x, y \rangle$$

with $\pi = \pi_1 \oplus \pi_2$ on $H_1 \oplus H_2$ and $x = x_1 \oplus x_2$, $y = y_1 \oplus y_2$.)

Observe that $\|f\|_{\ell_\infty(G)} = \sup_{t \in G} |f(t)| \le \|f\|_{B(G)}$. Clearly we have $\ell_1(G) \subset B(G)$ since the identity $\delta_s(t) = \langle \lambda(t)\delta_e, \delta_s \rangle$ shows that δ_s is in $B(G)$ with $\|\delta_s\|_{B(G)} \le 1$, hence for all f in $\ell_1(G)$

$$(2.3) \qquad \|f\|_{\ell_\infty(G)} \le \|f\|_{B(G)} \le \|f\|_{\ell_1(G)}.$$

More precisely, the identity $f(t) = \langle f, \delta_t \rangle = \langle f, \lambda(t)\delta_e \rangle$ shows that

$$\|f\|_{B(G)} \le \|f\|_{\ell_2(G)}.$$

Let G be a group and let $S \subset G$ be a semi-group included in G, i.e.

$$\forall s, t \in S \qquad st \in S.$$

We introduce the space $T_p(S)$ of all the functions $f \colon S \to \mathbb{C}$ which admit the following decomposition: there are functions $f_1 \colon S \times S \to \mathbb{C}$ and $f_2 \colon S \times S \to \mathbb{C}$ such that

$$f(st) = f_1(s,t) + f_2(s,t)$$

and

$$(2.4) \qquad \sup_{s \in S} \sum_{t \in S} |f_1(s,t)|^p < \infty, \quad \sup_{t \in S} \sum_{s \in S} |f_2(s,t)|^p < \infty.$$

We equip this space with the norm

$$\|f\|_{T_p(S)} = \inf \left\{ \sup_s \left(\sum_t |f_1(s,t)|^p \right)^{1/p} + \sup_t \left(\sum_s |f_2(s,t)|^p \right)^{1/p} \right\}$$

where the infimum runs over all such decompositions. We have clearly

$$\ell_p(S) \subset T_p(S) \subset \ell_\infty(S)$$

and $T_p(S)$ is a Banach space. The main result of this section is the following theorem from [BF2] (see below for complements).

Theorem 2.1. *If every uniformly bounded representation on a discrete group G is unitarizable, then*

$$T_1(G) \subset B(G).$$

Proof: Let $f \in T_1(G)$. Substituting s^{-1} for s in the definition, we find a decomposition of the form

(2.5) $$f(s^{-1}t) = a_1(s,t) + a_2(s,t)$$

and for some constant C we have

(2.6) $$\sup_s \sum_t |a_1(s,t)| \leq C \quad \sup_t \sum_s |a_2(s,t)| \leq C.$$

Let us denote by A_1 and A_2 the linear operators say from $\mathbb{C}^{(G)}$ into $\mathbb{C}^G$ admitting a_1 and a_2 as their representative matrices. Let $\check{f}$ be the function defined by $\check{f}(x) = f(x^{-1})$. Observe that $f(s^{-1}t)$ is nothing but the matrix of the operator $\rho(f)$ (of right convolution by $\check{f}$ from $\mathbb{C}^{(G)}$ to $\mathbb{C}^G$) defined by

$$\rho(f)g = g * \check{f}.$$

Hence we have

$$\rho(f) = A_1 + A_2,$$

and (2.6) can be rewritten as

(2.7) $$\|A_1\|_{B(\ell_\infty(G))} \leq C \quad \|A_2\|_{B(\ell_1(G))} \leq C.$$

By an obvious calculation (recall (2.1)) we have

$$\rho(f)\lambda(t) = \lambda(t)\rho(f) \qquad \forall t \in G.$$

Hence if we denote by $[a,b] = ab - ba$ the commutant of two linear operators a, b we have

$$0 = [\rho(f), \lambda(t)] = [A_1, \lambda(t)] + [A_2, \lambda(t)]$$

hence

(2.8) $$[A_1, \lambda(t)] = -[A_2, \lambda(t)].$$

Let us denote (for clarity)

$$D(a) = [A_2, a] = A_2 a - a A_2.$$

As is well known D is a derivation, i.e. we have (whenever it makes sense)

(2.9) $$D(ab) = D(a)b + aD(b).$$

It follows that if we set $H = \ell_2(G) \oplus \ell_2(G)$ the map $\pi(t) : H \to H$ defined by

$$\pi(t) = \begin{pmatrix} \lambda(t) & D(\lambda(t)) \\ 0 & \lambda(t) \end{pmatrix}$$

is a representation of G. Indeed, (2.9) immediately implies

$$\forall t, s \in G \qquad \pi(t)\pi(s) = \pi(ts).$$

Moreover, this representation is uniformly bounded. Indeed, by (2.7) and (2.8) the operator $[A_2, \lambda(t)] = D(\lambda(t))$ is bounded simultaneously on $\ell_\infty(G)$ and $\ell_1(G)$ with norm $\leq 2C$ (observe that $\lambda(t)$ is an isometry both on $\ell_1(G)$ and $\ell_\infty(G)$). By the Riesz-Thorin classical interpolation theorem (see also Proposition 2.15 in the appendix), it follows that

$$\|[A_2, \lambda(t)]\|_{B(\ell_2(G))} \leq 2C$$

and consequently

$$\|\pi(t)\| \leq 1 + 2C.$$

Finally, let $x = (0, \delta_e), y = (\delta_e, 0)$ and consider

$$\begin{aligned}
\langle \pi(t)x, y \rangle &= \langle D(\lambda(t))\delta_e \oplus \lambda(t)\delta_e, \delta_e \oplus 0 \rangle \\
&= \langle D(\lambda(t))\delta_e, \delta_e \rangle \\
&= \langle A_2\lambda(t)\delta_e - \lambda(t)A_2\delta_e, \delta_e \rangle \\
&= a_2(e, t) - a_2(t^{-1}, e).
\end{aligned}$$

Clearly if π is unitarizable, by the preceding equalities, then we have $\langle \pi(t)x, y \rangle = \langle \tilde{\pi}(t)Sx, S^{-1*}y \rangle$ for some $\tilde{\pi}$ unitary and some similarity S, hence the function $t \to \psi(t) = a_2(e, t) - a_2(t^{-1}, e)$ must be in $B(G)$. Finally we observe that by (2.5)

$$f(t) = a_1(e, t) + a_2(e, t)$$

hence

$$f(t) = a_1(e, t) + a_2(t^{-1}, e) + [a_2(e, t) - a_2(t^{-1}, e)]$$

and by (2.6) $t \to a_1(e, t) + a_2(t^{-1}, e)$ is in $\ell_1(G)$, hence a fortiori in $B(G)$, so then we conclude that $f \in B(G)$.

$\square$

We will denote by $\mathbb{F}_\infty$ (resp. $\mathbb{F}_N$) the free group with countably many generators (resp. with N generators). We will denote the generators by $g_1, g_2, \dots$. Let t be an element of $\mathbb{F}_\infty$ (resp. $\mathbb{F}_N$), i.e. t is a "word" written with the letters g_j and g_j^{-1}. The empty word is the unit element. When all possible cancellations have been made, the word t is called "reduced" and the total number of the remaining letters is called the "length" of t and is denoted by $|t|$. The length of the empty word is defined to be zero.

To obtain quickly an example let us immediately observe the following.

Lemma 2.2. Let f be the indicator function of the set of words of length 1 in $\mathbb{F}_\infty$. Then $f \in T_1(\mathbb{F}_\infty)$ but $f \notin B(\mathbb{F}_\infty)$.

Proof: Assume that $f \in B(\mathbb{F}_\infty)$ so that for some unitary representation $\pi \colon \mathbb{F}_\infty \to B(H)$ we have (2.2). Hence

$$(2.10) \qquad \langle \pi(g_j)x, y \rangle = \langle \pi(g_j)^*x, y \rangle = 1 \quad \text{and} \quad \langle \pi(t)x, y \rangle = 0 \text{ if } |t| \neq 1.$$

Let $a_j = \mathrm{Re}(\pi(g_j)) = \frac{1}{2}(\pi(g_j) + \pi(g_j)^*)$, note that $\|a_j\| \leq 1$ and define for each $n \geq 1$

$$(2.11) \qquad R = \prod_{j=1}^{n} \left(I + \frac{i}{\sqrt{n}} a_j \right).$$

Note that

$$\left\| I + \frac{i}{\sqrt{n}} a_j \right\|^2 = \|(I + i a_j n^{-1/2})^*(I + i a_j n^{-1/2})\|$$

$$= \|I + a_j^2/n\| \leq 1 + \frac{1}{n}.$$

Hence

$$\|R\|^2 \leq \left(1 + \frac{1}{n} \right)^n \leq e.$$

On the other hand, developing (2.11) we find

$$R = \frac{i}{2\sqrt{n}} \sum_{j=1}^{n} \pi(g_j) + \pi(g_j)^* + \sum_{|t| \neq 1} \psi(t)\pi(t)$$

for some finitely supported function ψ. Hence by (2.10)

$$\langle Rx, y \rangle = \frac{i}{2\sqrt{n}}(2n) + 0 = i\sqrt{n}.$$

Hence we obtain

$$\sqrt{n} = |\langle Rx, y \rangle| \leq \|R\| \, \|x\| \, \|y\| \leq \sqrt{e} \, \|x\| \, \|y\|$$

and letting $n \to \infty$ this yields the desired contradiction, which proves that $f \notin B(\mathbb{F}_\infty)$. Let us now check that $f \in T_1(\mathbb{F}_\infty)$. Let

$$f_1(s,t) = 1_{\{|st|=1, |s|>|t|\}}$$
$$f_2(s,t) = 1_{\{|st|=1, |s|<|t|\}}.$$

Then clearly

$$f(st) = f_1(s,t) + f_2(s,t).$$

We claim that

$$(2.12) \qquad \sup_s \sum_t |f_1(s,t)| \leq 1 \quad \text{and} \quad \sup_t \sum_s |f_2(s,t)| \leq 1.$$

Indeed, fix s in $\mathbb{F}_\infty$ and assume that st is a word of length one. This word of length one being a generator or its inverse it has to come either from s or from t. Assume now moreover that $|s| > |t|$. Then necessarily this generator comes from s and actually is the first letter of s, which we denote by g_s. Since $st = g_s$ we have $t = s^{-1}g_s$ hence t itself is determined by s. This shows the first half of (2.12), the other one is similar.   □

Corollary 2.3. *There are non-unitarizable uniformly bounded representations on* $\mathbb{F}_\infty$.

Remark. It is easy to modify the argument to construct for each $\varepsilon > 0$ a representation uniformly bounded by $1 + \varepsilon$ and still non-unitarizable.

We will use a very convenient amenability criterion due to Hulanicki (based on previous work of Kesten), as follows.

Theorem 2.4. *The following properties of a discrete group are equivalent.*

(i) *G is amenable.*

(ii) *There is a constant C such that for any finitely supported function f on G with $f \geq 0$ we have*

$$\sum_{t \in G} f(t) \leq C \Big\| \sum f(t) \lambda(t) \Big\|_{B(\ell_2(G))}.$$

(iii) *Same as (ii) with $C = 1$.*

Proof: Assume (i). Let φ be an invariant mean for G. Consider a net ψ^i in $\ell_1(G)$ with $\psi^i \geq 0$, $\|\psi^i\|_1 = 1$ and $\psi^i \to \varphi$ for the topology $\sigma(\ell_\infty(G)^*, \ell_\infty(G))$. Since φ is invariant, for any t in G $\delta_t * \psi^i - \psi^i \to 0$ for $\sigma(\ell_1(G), \ell_\infty(G))$. Passing to convex combinations of the ψ_i (and using the fact that the norm and weak closures of a convex set are the same) we obtain a net φ^i in $\ell_1(G)$ with $\varphi^i \geq 0$, $\|\varphi^i\|_1 = 1$ and $\|\delta_t * \varphi^i - \varphi^i\|_1 \to 0$ when $i \to \infty$. Now let $g^i(t) = (\varphi^i(t))^{1/2}$. Then $\|g^i\|_2 = 1$ and $\|\delta_t * g^i - g^i\|_2 \to 0$ when $i \to \infty$.
Now consider $f \geq 0$ and finitely supported. Assume that $\|\sum f(t)\lambda(t)\|_{B(\ell_2(G))} \leq 1$. Then

$$\left| \sum_{s,t} f(st^{-1}) g^i(t) \overline{g^i(s)} \right| \leq 1$$

hence

$$\left| \sum_{u,s} f(u) g^i(u^{-1}s) \overline{g^i(s)} \right| \leq 1.$$

Now since $\|\delta_u * g^i - g^i\|_2 \to 0$ when $i \to \infty$ we must have

$$\lim_{i \to \infty} \left| \sum_{u,s} f(u) g^i(s) \overline{g^i(s)} \right| \leq 1$$

hence $\left| \sum_u f(u) \right| \leq 1$. By homogeneity, this shows that (iii) and a fortiori (ii) hold.
We now show (ii) $\Rightarrow$ (iii). If (ii) holds for all convolution powers of f we have

$$\left(\sum f(t) \right)^n = \sum_{t \in G} \underbrace{f * \cdots * f}_{n \text{ times}}(t) \leq C \| \lambda(f * \cdots * f) \|_{B(\ell_2(G))}$$

$$= C \| \lambda(f)^n \|_{B(\ell_2(G))} \leq C \| \lambda(f) \|_{B(\ell_2(G))}^n$$

hence taking the n-th root and letting $n \to \infty$ we obtain (iii). Now, we show (iii) $\Rightarrow$ (i). Assume (iii).

Actually, we will use only that for any finite subset $E \subset G$ with $e \in E$ (e denotes the unit element of G) we have

$$|E| \le \left\| \sum_{t \in E} \lambda(t) \right\|_{B(\ell_2(G))}.$$

This implies that for each such E there is a sequence g_n in $\ell_2(G)$ such that $\|g_n\|_2 = 1$ and

$$\left\| \frac{1}{|E|} \sum_{t \in E} \lambda(t) g_n \right\|_2 = \left\| \frac{1}{|E|} \sum_{t \in E} \delta_t * g_n \right\|_2 \to 1$$

when $n \to \infty$. By the uniform convexity of $\ell_2(G)$ (recall that E is fixed) this implies that

$$\forall t \in E \qquad \|\delta_t * g_n - g_n\|_2 \to 0 \quad \text{when} \quad n \to \infty.$$

(Hint: to check this, simply compute

$$\sum_{s \in E} \left\| \frac{1}{|E|} \left(\sum_{t \in E} \delta_t * g_n \right) - \delta_s * g_n \right\|_2^2.)$$

Since this holds for all E, there must exist a net (g^i) in $\ell_2(G)$ such that $\|g^i\|_2 = 1$ and such that

(2.13) $\qquad \forall t \in G \qquad \|\delta_t * g^i - g^i\|_2 \to 0$

when $i \to \infty$.

Now for any f in $\ell_\infty(G)$ we can define

$$\varphi(f) = \lim_{\mathcal{U}} \langle f \cdot g^i, g^i \rangle$$

or equivalently

$$\varphi(f) = \lim_{\mathcal{U}} \sum_{s \in G} f(s) |g^i(s)|^2$$

where $\mathcal{U}$ is a ultrafilter finer than the above net. Clearly $\varphi(1) = 1 = \|\varphi\|$ and $\varphi \ge 0$. Moreover for each fixed t in G

$$\varphi(\delta_{t^{-1}} * f) = \lim_{\mathcal{U}} \sum_{s \in G} f(s) |g^i(t^{-1}s)|^2$$

$$= \lim_{\mathcal{U}} \langle f \cdot (\delta_t * g^i), \delta_t * g^i \rangle$$

hence by (2.13)

$$= \lim_{\mathcal{U}} \langle f \cdot g^i, g^i \rangle$$

$$= \varphi(f).$$

This shows that φ is an invariant mean, hence (i) holds. This completes the proof that (iii) $\Rightarrow$ (i). (As an exercise the reader can check that the same result remains true with the norm in $B(\ell_p(G))$ for any $1 < p < \infty$, but not for either $p = 1$ or $p = \infty$.) □

Using Theorem 2.1, we can deduce the following result from [W1].

Theorem 2.5. *Let* $1 \le p < \infty$. *A discrete group is amenable iff* $T_p(G) \subset \ell_p(G)$.

Proof: Assume G amenable. Let $f \in T_p(G)$. Let $f = f_1 + f_2$ be the corresponding decomposition satisfying (2.4). Observe that we have for any x in G

$$(2.14) \qquad f(st) = f_1(sx, x^{-1}t) + f_2(sx, x^{-1}t).$$

Hence considering the right side of (2.14) as a bounded function of x (for s, t fixed) and applying an invariant mean φ to it, we obtain

$$f(st) = \tilde{f}_1(s,t) + \tilde{f}_2(s,t)$$

where

$$(2.15) \qquad \tilde{f}_j(s,t) = \varphi[F_j(s,t)]$$

and where $F_j(s,t) \in \ell_\infty(G)$ is defined by $F_j(s,t)(x) = f_j(sx, x^{-1}t)$. Since φ is left invariant, we have

$$\forall y \in G \qquad \tilde{f}_j(s,t) = \tilde{f}_j(sy, y^{-1}t)$$

hence

$$\tilde{f}_j(s,t) = \tilde{f}_j(st, e) = \tilde{f}_j(e, st).$$

This implies

$$(2.16) \qquad f(x) = \tilde{f}_1(e, x) + \tilde{f}_2(x, e).$$

On the other hand, since φ is positive, by an obvious extension of Jensen's inequality to φ (Hint: for $p \ge 1$ the function $x \to |x|^p$ on $\mathbb{R}$ is the supremum of all the affine functions which it majorizes) we have

$$|\varphi(F)|^p \le \varphi(|F|^p) \quad \text{for any} \quad F \text{ in } \ell_\infty(G)$$

which allows us to deduce from (2.15) for any finite subset $E \subset G$

$$\forall s \in G \qquad \sum_{t \in E} |\tilde{f}_1(s,t)|^p \le \sup_x \sum_{t \in E} |f_1(sx, x^{-1}t)|^p \le \sup_x \sum_{t \in G} |f_1(sx, x^{-1}t)|^p$$

hence since E is arbitrary

$$\forall s \in G \qquad \sum_{t \in G} |\tilde{f}_1(s,t)|^p \le \sup_s \sum_{t \in G} |f_1(s,t)|^p,$$

and similarly

$$\forall t \in G \qquad \sum_{s \in G} |\tilde{f}_2(s,t)|^p \leq \sup_t \sum_{s \in G} |f_2(s,t)|^p.$$

By (2.16) this implies

$$\left(\sum |f(x)|^p \right)^{1/p} \leq \|f\|_{T_p(G)},$$

which proves the only if part.

The converse is easy by Theorem 2.4. Indeed, if $T_p(G) \subset \ell_p(G)$ we clearly have $T_1(G) \subset \ell_1(G)$ and by the closed graph theorem there is a constant C such that for all finitely supported functions $f \geq 0$ on G we have

$$\sum_{x \in G} f(x) \leq C\|f\|_{T_1(G)}.$$

Consider now finite subsets $E \subset G$, $F \subset G$. Let

$$v = \sum_{s \in G} f(s)\lambda(s).$$

We have

$$f * 1_F(s) = v(1_F)(s) = \sum_{x \in F} f(sx^{-1})$$

hence

$$|\langle 1_E, v1_F \rangle| = \left| \sum_{\substack{s \in E \\ x \in F}} f(sx^{-1}) \right| \leq \|v\|_{B(\ell_2(G))}(|E| \, |F|)^{1/2},$$

and consequently (change F to F^{-1})

$$\left| \sum_{(s,t) \in E \times F} f(st) \right| \leq \|v\|(|E| \, |F|)^{1/2}.$$

But it clearly follows from the next lemma (see also Proposition 2.16 in the appendix) that

$$\|f\|_{T_1(G)} \leq 2 \left\| \sum f(t)\lambda(t) \right\|_{B(\ell_2(G))},$$

hence by Theorem 2.4 we conclude that G is amenable.   □

Lemma 2.6. *[V1] Let S, T be arbitrary sets, and let $\{a(s,t) \mid s \in S, t \in T\}$ be complex numbers. Let*

$$A\colon \mathbb{C}^{(T)} \to \mathbb{C}^S$$

be the linear operator associated to the matrix (= kernel) $(a(s,t))$. (i.e. for any finitely supported function f on T we have $(Af)(s) = \sum_t a(s,t)f(t)$.)

(i) *Assume that for any n and for finite subsets $E \subset S$, $F \subset T$ with $|E| = |F| = n$ we have*

$$\sum_{(s,t)\in E\times F} |a(s,t)| \le n.$$

Then there is a decomposition $a(s,t) = a_1(s,t) + a_2(s,t)$ with

(2.17) $$\sup_s \sum_t |a_1(s,t)| \le 1 \quad \text{and} \quad \sup_t \sum_s |a_2(s,t)| \le 1.$$

(ii) *Conversely, if such a decomposition holds then we have for all E, F as in (i)*

$$\sum_{(s,t)\in E\times F} |a(s,t)| \le 2n.$$

Proof: We first prove (i). By a simple compactness argument, it suffices to establish this when S and T are finite sets, or equivalently when $|S| = |T| = N$. We will verify this by induction on the cardinality N of these sets. Let us assume that (i) has been established whenever $|S| = |T| = N-1$. Now consider sets S, T with $|S| = |T| = N$ and $\{a(s,t) \mid s \in S, t \in T\}$ satisfying the assumption in (i). Since $\frac{1}{N}\sum_{s\in S}\sum_{t\in T} |a(s,t)| \le 1$, there is an index s_0 such that $\sum_{t\in T} |a(s_0,t)| \le 1$. Similarly, there is an index t_0 such that $\sum_{s\in S} |a(s,t_0)| \le 1$.
Let S' (resp. T') be the sets with $|S'| = |T'| = N - 1$ obtained by removing s_0 (resp. t_0) from S and T respectively. Applying the induction hypothesis to $\{a(s,t) \mid s \in S', t \in T'\}$, we find a decomposition

$$\forall(s,t) \in S' \times T' \qquad a(s,t) = a_1(s,t) + a_2(s,t)$$

satisfying the bound (2.17). Then we extend a_1 and a_2 to $S \times T$ by setting

$$a_1(s_0,t) = a(s_0,t), \qquad a_2(s_0,t) = 0 \quad \forall t \in T'$$

and

$$a_2(s,t_0) = a(s,t_0), \qquad a_1(s,t_0) = 0 \quad \forall s \in S'.$$

By the initial choice of s_0 and t_0, this decomposition of $\{a(s,t) \mid (s,t) \in S \times T\}$ still satisfies (2.17). Thus we obtain (i) with $|S| = |T| = N$. By induction, this establishes the first part of Lemma 2.6. The second part is obvious.   □
A finite (resp. infinite) subset E of a discrete group G is called free if the subgroup it generates in G is isomorphic to the free group $\mathbb{F}_N$ (resp. $\mathbb{F}_\infty$) with the elements of E mapped to the generators of $\mathbb{F}_N$ (resp. $\mathbb{F}_\infty$) by the isomorphism. Equivalently, this means that E satisfies no nontrivial relation.
We now give a slight improvement of Lemma 2.2 and Corollary 2.3.

Lemma 2.7. *Let E be a free set in a discrete group G. Let $E^{-1} = \{t^{-1} \mid t \in E\}$.*

(i) *Then for any f supported by $E \cup E^{-1}$ we have*

(2.18) $$\left(\sum_{t\in E\cup E^{-1}} |f(t)|^2\right)^{1/2} \le 2\sqrt{e}\,\|f\|_{B(G)}.$$

(ii) If E is infinite, the indicator function of $E \cup E^{-1}$ is in $T_1(G)$ and not in $B(G)$. Moreover, for any subgroup $F \subset G$, $T_1(F)$ can be identified isometrically with the subspace of $T_1(G)$ of all the functions supported on F.

(iii) For each $2 \leq N < \infty$, IF_N contains an infinite free subset hence IF_∞ embeds in IF_N as a subgroup, $T_1(\mathrm{IF}_N) \not\subset B(\mathrm{IF}_N)$, and there are uniformly bounded non-unitarizable representations on IF_N.

Proof: (i) Assume $\|f\|_{B(G)} < 1$. Then there are H, a representation $\pi \colon G \to B(H)$ and $x, y \in H$ with $\|x\| < 1$, $\|y\| < 1$ such that

$$\forall t \in G \qquad f(t) = \langle \pi(t)x, y \rangle.$$

Let $(\alpha(t))_{t \in E \cup E^{-1}}$ be a finitely supported complex function such that $\sum |\alpha(t)|^2 \leq 1$. For (2.18) it suffices to show $\sum |\alpha(t)f(t)| \leq 2\sqrt{e}$. Let R in $B(H)$ be the sum of the following two products

$$R = (-i) \prod_{t \in E \cup E^{-1}} (1 + i \operatorname{Re}(\alpha(t)\pi(t))) + \prod_{t \in E \cup E^{-1}} (1 + i \operatorname{Im}(\alpha(t)\pi(t))).$$

Observe that since E is assumed free

$$R = \sum_{t \in E \cup E^{-1}} \alpha(t)\pi(t) + \sum_{t \notin E \cup E^{-1}} \psi(t)\pi(t)$$

for some finitely supported function ψ. Since $f(t) = \langle \pi(t)x, y \rangle$ is assumed supported by $E \cup E^{-1}$, we have

$$\langle Rx, y \rangle = \sum_{t \in E \cup E^{-1}} \alpha(t)f(t)$$

hence $\left| \sum_{t \in E \cup E^{-1}} \alpha(t)f(t) \right| < \|R\|$. But clearly (as in Lemma 2.2 above)

$$\|R\| \leq 2 \prod_{t \in E \cup E^{-1}} (1 + \|\alpha(t)\pi(t)\|^2)^{1/2}$$

$$\leq 2 \prod (1 + |\alpha(t)|^2)^{1/2}$$

$$\leq 2 \exp \frac{1}{2} \sum |\alpha(t)|^2$$

$$\leq 2\sqrt{e}.$$

This proves (2.18). We now check (ii). Clearly $1_{E \cup E^{-1}}$ is not in $B(G)$ if E is infinite by (2.18). Let us check that it is in $T_1(G)$. Clearly by Lemma 2.2 and the obvious identification between IF_∞ and the subgroup F generated by E in G, we know that $1_{E \cup E^{-1}} \notin T_1(F)$, hence it suffices to show that $T_1(F)$ can be identified isometrically with $T_1(G) \cap \{f \mid \operatorname{supp}(f) \subset F\}$. To check that, let $G = \bigcup_m Ft_m$ be a partition of G into disjoint right cosets relative to the subgroup F. Observe that if $st \in F$ with $s \in Ft_m$, $t^{-1} \in Ft_{m'}$ then necessarily $m = m'$. Now consider f in $T_1(G)$ supported by F. We have obviously

$$\|f_{|F}\|_{T_1(F)} \le \|f\|_{T_1(G)}.$$

Conversely, assume that $\|f_{|F}\|_{T_1(F)} < 1$. We have a decomposition

$$\forall s, t \in F \qquad f(st) = f_1(s,t) + f_2(s,t)$$

with

$$\sup_{s \in F} \sum_{t \in F} |f_1(s,t)| + \sup_{t \in F} \sum_{s \in F} |f_2(s,t)| < 1.$$

We now extend f_1 and f_2 to $G \times G$ as follows: if $s \in Ft_m$, $t^{-1} \in Ft_m$ we have $st_m^{-1} \in F$, $t_m t \in F$ and we may define

$$\tilde{f}_1(s,t) = f_1(st_m^{-1}, t_m t)$$

and

$$\tilde{f}_2(s,t) = f_2(st_m^{-1}, t_m t).$$

If $s \in Ft_m$ and $t^{-1} \in Ft_{m'}$ with $m \ne m'$ we set $\tilde{f}_1(s,t) = \tilde{f}_2(s,t) = 0$. Then we have $f(st) = \tilde{f}_1(s,t) + \tilde{f}_2(s,t)$ for all s, t in G and obviously

$$\sup_{s \in G} \sum_{t \in G} |\tilde{f}_1(s,t)| = \sup_{s \in F} \sum_{t \in F} |f_1(s,t)|$$

and similarly for f_2, so that we conclude as announced that $\|f\|_{T_1(G)} < 1$. In conclusion we have established $\|f\|_{T_1(G)} = \|f_{|F}\|_{T_1(\mathbb{F})}$ for all f supported by F, which completes the proof of (ii).

To check (iii) it now suffices by the preceding results to show that if $2 \le N < \infty$, $\mathbb{F}_N$ contains $\mathbb{F}_\infty$ as a subgroup, or equivalently, we claim that $\mathbb{F}_N$ contains an infinite free set. Indeed, let a, b be two distinct generators among the generators of $\mathbb{F}_N$, then the sequence $\{a^n b^n, n > 0\}$ clearly is an infinite free subset of $\mathbb{F}_N$. (An other example: if x and y are any two noncommuting elements in a free group, the set $\{x^{-n} y x^n, n > 0\}$ is free.) This proves our claim and completes the proof. □

We denote by $C^*(G)$ the C^*-algebra obtained by completing $\ell_1(G)$ with respect to the C^*-norm defined as follows

$$\forall f \in \ell_1(G) \qquad \|f\|_{C^*(G)} = \sup \left\| \sum f(t)\pi(t) \right\|$$

where the supremum runs over all unitary representations of G.

It is very easy to check with this notation that $B(G)$ can be identified with the dual space of $C^*(G)$ the duality being the usual one

$$\forall f \in \ell_1(G) \quad \forall g \in B(G) \qquad \langle g, f \rangle = \sum_{t \in G} g(t) f(t).$$

To summarize, we note the isometric identity

$$B(G) = (C^*(G))^*.$$

Remark 2.8. Actually the conclusion of Theorem 2.1 can be strengthened, if every uniformly bounded representation on a discrete group G is unitarizable, then $T_1(G) \subset \ell_2(G)$. This follows from the fact (proved originally in [TJ]) that the dual of a C^*-algebra is of cotype 2, i.e. there is a constant C such that for any finite sequence $x_1, \ldots, x_n$ in the dual A^* of a C^*-algebra A we have

$$(2.19) \qquad \left(\sum_1^n \|x_i\|^2 \right)^{1/2} \leq C \left(\int \left\| \sum \varepsilon_i x_i \right\|^2 d\mu \right)^{1/2}$$

where the average on the right side is the uniform average over the 2^n choices of signs $\varepsilon_1 = \pm 1, \ldots, \varepsilon_n = \pm 1$. Since the norm of a function f in $T_1(G)$ depends only on $|f|$, this inequality (2.19) shows that the inclusion $T_1(G) \subset B(G)$ proved in Theorem 2.1 implies automatically $T_1(G) \subset \ell_2(G)$. Indeed, we have for all finitely supported f in $T_1(G)$

$$\left(\sum \|f(t)\delta_t\|_{B(G)}^2 \right)^{1/2} \leq C \sup_{\varepsilon_t \pm 1} \left\| \sum \varepsilon_t f(t)\delta_t \right\|_{B(G)}$$

hence by Theorem 2.1

$$\leq CC' \sup \left\| \sum \varepsilon_t f(t)\delta_t \right\|_{T_1(G)}$$
$$= CC'\|f\|_{T_1(G)},$$

and since $\|\delta_t\|_{B(G)} \geq \|\delta_t\|_{\ell_\infty(G)} = 1$ we obtain

$$\left(\sum |f(t)|^2 \right)^{1/2} \leq CC'\|f\|_{T_1(G)}$$

as announced.

We will now give (following [B2]) a simple modification of the preceding construction to produce examples of "power bounded" operators (i.e. operators T such that $\sup_{n \geq 0} \|T^n\| < \infty$) which are not polynomially bounded.

Theorem 2.9. For any f in $T_1(\mathbb{N})$ there is a power bounded operator T on a Hilbert space H and there are x, y in H such that

$$\forall n \geq 0 \qquad f(n) = \langle T^n x, y \rangle.$$

More generally, let G be any Abelian group, consider a semi-group $S \subset G$, containing the unit element. Then for any f in $T_1(S)$ there is a uniformly bounded representation $\pi \colon S \to B(H)$ on some Hilbert space H (i.e. a map satisfying $\pi(ts) = \pi(t)\pi(s)$ for all t, s in S and $\sup_{t \in S} \|\pi(t)\| < \infty$) for which there are elements x, y in H such that

$$\forall t \in S \qquad f(t) = \langle \pi(t)x, y \rangle.$$

Proof: The first part is a particular case of the second one with $S = \mathbb{N}$ and $G = \mathbb{Z}$, hence it suffices to prove the second one. We will view the space $\ell_2(S)$

as embedded into $\ell_2(G)$ in the usual way (any function in $\ell_2(\mathcal{S})$ is viewed as a function on G which is zero outside $\mathcal{S}$). For each t in $\mathcal{S}$, let us denote by $S(t)$ the operator (analogous to the shift operator) defined on $\ell_2(\mathcal{S})$ by

$$(S(t)g)(s) = g(t^{-1}s)1_{\mathcal{S}}(t^{-1}s).$$

Then the operator $S(t)^*\colon \ell_2(\mathcal{S}) \to \ell_2(\mathcal{S})$ satisfies

$$(S(t)^*g)(s) = g(st) \qquad \forall s \in \mathcal{S}.$$

Clearly we have $S(t)^*S(s)^* = S(ts)^*$, and also of course $S(ts) = S(t)S(s)$ (recall that G is assumed commutative).

Now consider the operator $h_f\colon \ell_2(\mathcal{S}) \to \ell_2(\mathcal{S})$ which admits $(f(st))_{s,t\in\mathcal{S}}$ as its representative matrix. In analogy with the classical identities satisfied by Hankel operators, we have

(2.20) $$\qquad \forall t \in \mathcal{S} \qquad h_f S(t) = S(t)^* h_f.$$

Indeed for any g in $\ell_2(\mathcal{S})$ we have

$$[h_f S(t)g](s) = [S(t)^* h_f g](s) = \sum_{u \in \mathcal{S}} f(stu)g(u).$$

Now assume that $f \in T_1(\mathcal{S})$. As in the proof of Theorem 2.1 this means that we have a decomposition $f(st) = a_1(s,t) + a_2(s,t)$ or equivalently

$$h_f = A_1 + A_2$$

such that for some constant C, $\sup\limits_{s} \sum\limits_{t} |a_1(s,t)| \le C$ and $\sup\limits_{t} \sum\limits_{s} |a_2(s,t)| \le C$ or equivalently

$$\|A_1\|_{B(\ell_\infty(\mathcal{S}))} \le C \quad \text{and} \quad \|A_2\|_{B(\ell_1(\mathcal{S}))} \le C.$$

By (2.20) we have

$$S(t)^* A_1 - A_1 S(t) = -(S(t)^* A_2 - A_2 S(t)).$$

Since $S(t)$ and $S(t)^*$ are clearly of norm ≤ 1 both on $\ell_1(\mathcal{S})$ and $\ell_\infty(\mathcal{S})$ we have therefore again by interpolation

$$\|S(t)^* A_2 - A_2 S(t)\|_{B(\ell_2(\mathcal{S}))} \le 2C.$$

For all T in $B(\ell_2(\mathcal{S}))$ we define

$$\delta(T) = A_2 T - T^* A_2$$

Let $H' = \ell_2(\mathcal{S}) \oplus \ell_2(\mathcal{S})$ and let $\pi'(t)$ be the map defined on H' by

$$\pi'(t) = \begin{pmatrix} S(t)^* & \delta(S(t)) \\ 0 & S(t) \end{pmatrix}.$$

Observe that if T_1, T_2 are commuting operators we have

$$\delta(T_1 T_2) = T_1^* \delta(T_2) + \delta(T_1) T_2.$$

Hence since G is assumed commutative, we have $\pi'(s)\pi'(t) = \pi'(st)$ for all s, t in S. Moreover, we have

$$\|\pi'(t)\| \leq 1 + 2C.$$

Finally, let e be the unit element of S and let δ_e be the corresponding basis vector of $\ell_2(S)$ (i.e. $\delta_e(s) = 1$ iff $s = e$) and let $x' = (0, \delta_e)$, $y' = (\delta_e, 0)$ in H'. We have

$$\langle \pi'(t)x', y' \rangle = a_2(e, t) - a_2(t, e)$$

hence again

$$f(t) = f(et) = [a_2(e, t) - a_2(t, e)] + a_2(t, e) + a_1(e, t).$$

Since $\sum |a_2(t, e)| < \infty$, $\sum |a_1(e, t)| < \infty$, there is a uniformly bounded representation $\pi'' \colon S \to B(H'')$ and x'', y'' in H'' such that

$$a_2(t, e) + a_1(e, t) = \langle \pi''(t)x'', y'' \rangle,$$

hence if we set $\pi = \pi' \oplus \pi''$ and $x = (x', x'')$, $y = (y', y'')$ we obtain

$$\forall t \in S \qquad f(t) = \langle \pi(t)x, y \rangle.$$

□

Let S be as above a semi-group sitting in a commutative group G. Let $B(S)$ be the set of functions f on S which are the coefficients of a contractive homomorphism $\pi \colon S \to B(H)$ (i.e. f is of the form $f(t) = \langle \pi(t)x, y \rangle$ and $\|\pi(t)\| \leq 1$). The preceding result shows that if $T_1(S) \not\subset B(S)$ then there are uniformly bounded homomorphisms on S which are not similar to contractive homomorphisms.

We will now show that this is the case if $S = \mathbb{N}$, $G = \mathbb{Z}$. Actually we will even obtain finally a stronger statement (see Corollary 2.14 below). Of course in the case $S = \mathbb{N}$, every homomorphism $\pi \colon \mathbb{N} \to B(H)$ is of the form $\pi(n) = T^n$ for some T in $B(H)$, and π is uniformly bounded (resp. contractive) iff T is power bounded (resp. a contraction).

We will need some Fourier analysis relative to the space H_∞. Recall that this space can be defined as follows. We first denote simply by L_∞ the space $L_\infty(\mathbf{T}, m)$ relative to the unidimensional torus equipped with its normalized Haar measure m. Then we may define H_∞ as the subspace of L_∞ formed of all the functions f with Fourier transform $\hat{f}$ vanishing on the negative integers. Recall that this coincides with the space of all (non-tangential) boundary values of bounded analytic functions on the interior of the unit disc. Recall also that, just as L_∞ is the dual of L_1, H_∞ is naturally a dual space, namely it is the dual of the quotient L_1/H_1^0, where H_1^0 is the pre-annihilator of H_∞ (see e. g. [R4]).

We will use the following classical fact, which we prove following [Fou].

Lemma 2.10. *Let $\{n_k, k \geq 0\}$ be a sequence of positive integers such that $n_{k+1} > 2n_k$. Then for any sequence of complex numbers (α_k) such that $\sum |\alpha_k|^2 \leq 1$ there is a function F in H_∞ with $\|F\|_{H_\infty} \leq \sqrt{e}$ and with Fourier coefficients $\{\widehat{F}(n) \mid n \geq 0\}$ satisfying*

$$\forall k \geq 0 \qquad \widehat{F}(n_k) = \alpha_k.$$

Moreover, if the sequence (α_k) is finitely supported there is a polynomial F with these properties.

Proof: The proof is based on the identity

$$(2.21) \qquad \forall a, b \in \mathbb{C} \qquad |a + \xi b|^2 + |b - \bar{\xi} a|^2 = (1 + |\xi|^2)(|a|^2 + |b|^2).$$

Then we define inductively a sequence of polynomials

$$P_0(z) = \alpha_0 z^{n_0} \qquad Q_0(z) \equiv 1$$

and for $k > 0$

$$P_k(z) = P_{k-1}(z) + \alpha_k z^{n_k} Q_{k-1}(z)$$
$$Q_k(z) = Q_{k-1}(z) - \bar{\alpha}_k \bar{z}^{n_k} P_{k-1}(z).$$

Then by (2.21) we have

$$\forall z \in \partial D \qquad |P_k(z)|^2 + |Q_k(z)|^2 = (1 + |\alpha_k|^2)(|P_{k-1}(z)|^2 + |Q_{k-1}(z)|^2)$$

hence since $|P_0(z)|^2 + |Q_0(z)|^2 = 1 + |\alpha_0|^2$ we have

$$|P_k(z)|^2 + |Q_k(z)|^2 \leq \prod_{j=0}^{k} (1 + |\alpha_j|^2) \leq e$$

Note that by induction (using $n_k \geq n_{k-1}$) it can be checked that for all $k \geq 0$

$$\text{supp } \widehat{P}_k \subset [0, n_k], \quad \text{supp } \widehat{Q}_k \subset [-n_k, 0], \quad \text{and } \widehat{Q}_k(0) = 1.$$

We leave this as an exercise to the reader. In particular $P_k \in H_\infty$ and $\|P_k\|_{H_\infty} \leq \sqrt{e}$ for all k. Let F be a weak-$*$ limit point of the sequence $\{P_k\}$ in H_∞. We claim that F has the desired property.
Indeed, since supp $(\widehat{P}_{k-1}) \subset [0, n_{k-1}]$ we have $\widehat{P}_k(n_k) = \alpha_k \widehat{Q}_{k-1}(0) = \alpha_k$, and moreover (using $n_k > 2n_{k-1}$) for any $0 \leq j < k$

$$\widehat{P}_k(n_j) = \widehat{P}_{k-1}(n_j) = \cdots = \widehat{P}_j(n_j) = \alpha_j.$$

Therefore for any j we have in the limit $\widehat{F}(n_j) = \alpha_j$. □

Remark 2.11. It is interesting to observe that the same argument yields the following extension. Let M be a von Neumann algebra, with predual M_*. We will denote by $H^\infty(M)$ the space of bounded M-valued analytic functions $f: D \to M$, equipped with the norm $\sup_{z \in D} \|f(z)\|_M$. Here again the Fourier transform of the non-tangential boundary values (which exist only in the strong operator topology) makes sense for all n in $\mathbb{Z}$: for any x in M_* and $z \in D$, we let

$f_x(z) = < f(z), x >$ so that $f_x \in H_\infty$, and we set $< \hat{f}(n), x > = \hat{f}_x(n)$. This defines $\hat{f}(n)$ as an element of $(M_*)^* = M$. In this extended setting, one can still show that $H^\infty(M)$ is naturally a dual space. See e. g. [RR, p. 81] for complementary information.

Now let $\alpha_k \in M$ be such that $\sum \|\alpha_k\|^2 \leq 1$. Then there is a function f in $H_\infty(M)$ such that $\hat{f}(n_k) = \alpha_k$ for all $k \geq 0$ and such that $\|f\|_{H_\infty(M)} \leq \sqrt{e}$. To show this we use the following identity valid for all a, b, ξ in M

$$(a + \xi b)^*(a + \xi b) + (b - \xi^* a)^*(b - \xi^* a) = a^*(1 + \xi\xi^*)a + b^*(1 + \xi^*\xi)b$$
$$\leq (1 + \|\xi\|^2)(a^*a + b^*b).$$

Then we define again $P_k(z)$ and $Q_k(z)$ as above for $|z| = 1$. We now have $P_0^* P_0 + Q_0^* Q_0 \leq 1 + \|\alpha_0\|^2$ and

$$\forall k \geq 1 \qquad P_k^* P_k + Q_k^* Q_k \leq (1 + \|\alpha_k\|^2)(P_{k-1}^* P_{k-1} + Q_{k-1}^* Q_{k-1})$$

hence

$$P_k^* P_k + Q_k^* Q_k \leq \prod_{j=0}^{k} (1 + \|\alpha_j\|^2) \leq e$$

so that $\|P_k(z)\| \leq \sqrt{e}$ for all z in ∂D. The rest of the argument is the same.

Conjecture. *There is a constant C such that if a sequence (α_k) in M satisfies*

$$(2.22) \qquad \max\left\{ \left\|\sum \alpha_k^* \alpha_k \right\|, \left\|\sum \alpha_k \alpha_k^* \right\| \right\} \leq 1$$

where the convergence is meant in the strong operator topology, then there is a function f in $H_\infty(M)$ such that

$$(2.23) \qquad \hat{f}(n_k) = \alpha_k \quad and \quad \|f\|_{H_\infty(M)} \leq C.$$

(Note that the converse statement is obvious, if (2.23) holds with $C = 1$ then (2.22) holds.)

As a consequence of Lemma 2.10, we have

Proposition 2.12. *Consider a function $f \colon \mathbb{N} \to \mathbb{C}$ for which there is a polynomially bounded operator T on a Hilbert space H and x, y in H such that*

$$\forall n \geq 0 \qquad f(n) = \langle T^n x, y \rangle.$$

If f is supported by a sequence $\{n_k\}$ as above then $\sum_{k \geq 0} |f(n_k)|^2 < \infty$.

Proof: Since T is polynomially bounded there is a constant C such that $\|P(T)\| \leq C\|P\|_\infty$ for all polynomials P. Let α_k be finitely supported with $\sum |\alpha_k|^2 \leq 1$ and let F be the polynomial given by Lemma 2.10. We have

$$\langle F(T)x, y \rangle = \sum \hat{F}(n)\langle T^n x, y \rangle = \sum \hat{F}(n_k)\langle T^{n_k} x, y \rangle = \sum \alpha_k f(n_k)$$

hence

$$\left|\sum \alpha_k f(n_k)\right| \leq \|F(T)\| \|x\| \|y\| \leq C\|F\|_\infty \|x\| \|y\|,$$

so that we conclude that

$$\left(\sum |f(n_k)|^2\right)^{1/2} \leq C\sqrt{e}\, \|x\| \|y\|.$$

$\square$

Lemma 2.13. *Let $\{n_k\}$ be a sequence of integers as above such that $n_k > 2n_{k-1}$. Then the indicator function of the sequence $\{n_k\}$ is in $T_1(\mathbb{N})$.*

Proof: Let $E = \{n_k\}$. Observe that for any integer $n \geq 0$ we have

$$(2.24) \qquad\qquad |E \cap [n, 2n]| \leq 1.$$

Consider then the decomposition

$$\forall i, j \in \mathbb{N} \qquad 1_E(i+j) = 1_{\{i+j\in E, i\geq j\}} + 1_{\{i+j\in E, i<j\}}.$$

Note that if i is given $i + j \in E$ and $i \geq j$ imply $i \leq i + j \leq 2i$ hence by (2.24) we have

$$\sup_i \sum_j 1_{\{i+j\in E, i\geq j\}} \leq 1.$$

Similarly

$$\sup_j \sum_i 1_{\{i+j\in E, i<j\}} \leq 1,$$

so that $\|1_E\|_{T_1(\mathbb{N})} \leq 2$. $\square$

We can now reach our goal. By combining Lemma 2.13, Proposition 2.12 and Theorem 2.9, we obtain immediately.

Corollary 2.14. *There is a power bounded operator T on a (separable) Hilbert space which is not polynomially bounded (hence a fortiori not similar to a contraction).*

The first example of such an operator is due to Foguel [Fo] (and Lebow [Le] showed that Foguel's example is not polynomially bounded). We have followed the approach of Bożejko [B2]. (See also [Pe1].)

Appendix. In this appendix, we give a direct proof (not using interpolation) of the fact that an operator A which has simultaneously norm ≤ 1 on ℓ_1 and on ℓ_∞ must have norm ≤ 1 on ℓ_2. Equivalently we have

Proposition 2.15. *Let (a_{ij}) be an infinite matrix such that $\sup_i \sum_j |a_{ij}| \leq 1$ and $\sup_j \sum_i |a_{ij}| \leq 1$. Then (a_{ij}) defines an operator A: $\ell_2 \to \ell_2$ with $\|A\| \leq 1$.*

Proof: The following simple proof goes back to Schur.
Let x, y be arbitrary in the unit ball of ℓ_2. We can write

$$\sum_{ij} |a_{ij}| |x_i| |y_j| = \sum_i |x_i| \sum_j |a_{ij}|^{1/2} \cdot (|a_{ij}|^{1/2} |y_j|)$$

hence by Cauchy-Schwarz

$$\leq \sum_i |x_i| \left(\sum_j |a_{ij}| \right)^{1/2} \left(\sum_j |a_{ij}| |y_j|^2 \right)^{1/2} \leq \sum_i |x_i| \left(\sum_j |a_{ij}| |y_j|^2 \right)^{1/2}$$

and by Cauchy-Schwarz again

$$\leq \left(\sum_i |x_i|^2 \right)^{1/2} \left(\sum_{ij} |a_{ij}| |y_j|^2 \right)^{1/2} \leq \|x\| \|y\| \sup_j \sum_i |a_{ij}| \leq \|x\| \|y\|.$$

A fortiori this yields $|\langle Ax, y \rangle| \leq \|x\| \|y\|$ so that we obtain $\|A\| \leq 1$ as announced.
$\square$

We wish to record here the following

Proposition 2.16. *Let $A = (a_{ij})$ be a bounded operator on ℓ_2 with $a_{ij} \geq 0$. Then for all matrices (b_{ij}) we have $|\sum a_{ij} b_{ij}| \leq \|A\| \max \left\{ \sum_i \sup_j |b_{ij}|, \right.$*
$\left. \sum_j \sup_i |b_{ij}| \right\}$. *Consequently, if $\|A\| < 1$ there is a decomposition $a_{ij} = \alpha_{ij} + \beta_{ij}$ where (α_{ij}), (β_{ij}) are scalars such that*

$$\sup_i \sum_j |\alpha_{ij}| + \sup_j \sum_i |\beta_{ij}| \leq 1.$$

Proof: Let $x_i = \sup_j |b_{ij}|^{1/2}$ and $y_j = \sup_i |b_{ij}|^{1/2}$. Then

$$\left| \sum a_{ij} b_{ij} \right| \leq \sum_{ij} a_{ij} x_i y_j \leq \|A\| \max \{ \sum_i |x_i|^2, \sum_j |y_j|^2 \},$$

which yields the first inequality. The second part follows immediately by duality, (using the fact that the dual of $X_0 \cap X_1$ is $X_0^* + X_1^*$, applied with $X_0 = \ell_1(c_0)$ and X_1 the same space but transposing i and j.)
$\square$

Remark. The classical criterion of Schur stated above as Proposition 2.15 is optimal in some sense: if an operator $A: \ell_2 \to \ell_2$ with $\|A\| \leq 1$ has an associated matrix (a_{ij}) with non-negative entries then (a_{ij}) admits a decomposition

$$a_{ij} = |a_{ij}^0|^{1/2} |a_{ij}^1|^{1/2}$$

with

$$\sup_i \sum_j |a_{ij}| \leq 1 \quad \text{and} \quad \sup_j \sum_i |a_{ij}| \leq 1.$$

For a proof of this observation and its interpretation in terms of the complex interpolation method, see [P6].

Notes and Remarks on Chapter 2

We start with some complements on the space $B(G)$. To simplify, we assume the group G discrete. Recall that a function $F : G \to \mathbb{C}$ is called positive definite if for every finitely supported function $\alpha : G \to \mathbb{C}$, we have

$$\sum_{s,t \in G} \alpha(s)\overline{\alpha(t)}F(s^{-1}t) \geq 0.$$

By a well known result of Bochner, a function f is positive definite iff there is a unitary representation $\pi : G \to B(H)$ and an element $x \in H$ such that

$$\forall t \in G \quad f(t) = \langle \pi(t)x, x \rangle.$$

Then, every function f in $B(G)$ can be written as $f = f_1 - f_2 + i(f_3 - f_4)$, with $f_1, ..., f_4$ all positive definite functions on G. Indeed, this is easy to deduce from (2.2) and the polarisation formula for sesquilinear forms. As a corollary, a function f is in $B(G)$ iff it is a linear combination of positive definite functions. The order structure induced on $B(G)$ by the cone of positive definite functions coincides with its ordering as the dual of the C^*-algebra $C^*(G)$. The states on $C^*(G)$ correspond exactly to the positive definite functions f taking the value one on the unit element.

There are numerous known characterizations of amenable groups ([Gr, Pat, Pi]). The well known criterion given above as Theorem 2.4 (due to Hulanicki) originates in Kesten's thesis devoted to random walks on discrete groups such as the free group. The above proof uses an idea of Day and works equally well if we replace the norm in $B(\ell_2(G))$ by that of $B(\ell_p(G))$ with $1 < p < \infty$. The key point is the uniform convexity of $\ell_p(G)$. See [AO] for related information.

The same criterion has a nice reformulation involving the full C^*-algebra $C^*(G)$ that was defined before Remark 2.8. To state this, we need to recall a classical notation: we denote by $C^*_\lambda(G)$ the C^*-algebra generated in $B(\ell_2(G))$ by $\{\lambda(t) \mid t \in G\}$. Note that $C^*_\lambda(G)$ coincides with the closed linear span of $\{\lambda(t) \mid t \in G\}$ in $B(\ell_2(G))$.

In general there is clearly a (C^*-algebraic) representation $Q: C^*(G) \to C^*_\lambda(G)$ which is onto, so that $C^*_\lambda(G)$ can be viewed as a quotient of $C^*(G)$ by an ideal $I = \ker(Q)$. Then the criterion alluded to is simply that G is amenable iff $C^*_\lambda(G)$ can be identified with $C^*(G)$, i.e. iff Q is injective.

Let

$$B_\lambda(G) = C^*_\lambda(G)^*.$$

Equivalently, $B_\lambda(G) \subset B(G)$ is the orthogonal of $\ker(Q)$. Therefore, the identity $B_\lambda(G) = B(G)$ is equivalent to $\ker(Q) = \{0\}$, hence characterizes amenable groups.

Note that since $B_\lambda(G) \subset B(G)$, $B_\lambda(G)$ is formed of certain coefficients of unitary representations of G. In particular, let $A(G)$ be the space of all coefficients of the single representation λ, $i.e.$ the set of all functions $f : G \to \mathbb{C}$ of the form $f = g * h$ with g, $h \in L_2(G)$ ($i.e.$ $\ell_2(G)$ in the discrete case). While it is by no means obvious, it turns out this is a linear space, and even a Banach algebra for the pointwise multiplication when equipped with the norm $\|f\| = \inf\{\|g\|_2 \|h\|_2\}$. It is an easy exercise to check that $A(G) \subset B_\lambda(G)$. Moreover, this inclusion is an isometric embedding. The dual $A(G)^*$ can be identified isometrically with the von Neumann algebra generated by $C_\lambda^*(G)$. The space $A(G)$ is usually referred to as the "Fourier algebra" of G.

We refer to Eymard's classical thesis [Ey] and to the books [Gr, Pat, Pi] for more informations.

Let $c > 1$ be a constant. We may denote by $B_c(G)$ the space of all coefficients of the representations of G which are uniformly bounded by c; more precisely, a function $f \colon G \to \mathbb{C}$ belongs to $B_c(G)$ if there is a uniformly bounded representation $\pi \colon G \to B(H_\pi)$ with $\sup_{t \in G} \|\pi(t)\| \le c$ and vectors x, y in the Hilbert space H_π such that

$$(2.25) \qquad\qquad f(t) = \langle \pi(t)x, y \rangle.$$

We define

$$\|f\|_{B_c(G)} = \inf\{\|x\| \|y\|\}$$

where the infimum runs over all possible x, y, π such that (2.25) holds. It is not hard to check that $B_c(G)$ is a Banach space and that if $f \in B_c(G)$ and $g \in B_d(G)$ then the pointwise product fg is in $B_{cd}(G)$. It would be interesting to have a nice description of the functions in the space $\bigcup_{c>1} B_c(G)$. For some time, there was some speculation that the latter space was the same as the space $M_0(G)$ of Herz-Schur multipliers on G. This space is defined as follows: a function $f \colon G \to \mathbb{C}$ belongs to $M_0(G)$ if the operator $T_f \colon B(\ell_2(G)) \to B(\ell_2(G))$ defined by $T_f((a(s,t))) = (f(st^{-1})a(s,t))$ is bounded on $B(\ell_2(G))$. We define

$$\|f\|_{M_0(G)} = \|T_f\|_{B(\ell_2(G)) \to B(\ell_2(G))}.$$

The linear transformation T_f is what is called a Schur multiplier. These multipliers are studied in more detail in Chapter 6 (see Theorem 6.3).

It is easy to see, using Theorem 5.1 below, that

$$(2.26) \qquad\qquad \bigcup_{c>1} B_c(G) \subset M_0(G)$$

and the converse inclusion remained open until Haagerup [H6] disproved it.

It was recently proved in [P3] that the space $T_2(G)$ (defined before Theorem 2.1) coincides with the space of all functions $f \colon G \to \mathbb{C}$ such that the pointwise product $\epsilon.f$ is in $M_0(G)$ for any bounded function $\epsilon\colon G \to \mathbb{C}$. In other words, $T_2(G)$ is the "unconditional hull" of $M_0(G)$. With $\mathbb{N}$ in the place of G, these results were previously obtained in [B2].

A function on the free group is called "radial" if its value on an element (= a word) depends only on the length of the corresponding word. The radial Herz-Schur multipliers on the free group with finitely many generators are described in [W3].

There are many known characterizations of amenability in terms of the coincidence of two function spaces or two spaces of multipliers on G, see [Pi]. However, the following seems to be still open:
Let $M(G)$ be the space of functions $f\colon G \to \mathbb{C}$ such that the above map T_f is bounded when restricted to the reduced C^*-algebra $C_\lambda^*(G)$. We equip this space with the obvious norm

$$\|f\|_{M(G)} = \|T_f|_{C_\lambda^*(G)}\|_{C_\lambda^*(G) \to C_\lambda^*(G)}.$$

This is the same as the space of bounded multipliers of the Fourier algebra $A(G)$. Note that obviously $M_0(G) \subset M(G)$. Hence we have in general (recall (2.26)) the inclusions

$$A(G) \subset B_\lambda(G) \subset B(G) \subset \bigcup_{c>1} B_c(G) \subset M_0(G) \subset M(G).$$

It is apparently unknown whether the identity $M(G) = M_0(G)$ characterizes amenable groups, see [JoV] for some results on this question. On the other hand, it is known (cf. [B3, W1]) that, in the discrete case, $B(G) = M_0(G)$ iff G is amenable. This improves on a previous result (see [Ne]) showing that $B(G) = M(G)$ iff G is amenable. Note however, that this latter result is known (see [Los]) for general locally compact groups.
Whether the equality $B(G) = \bigcup_{c>1} B_c(G)$ characterizes amenability is open, this seems very close to the Revised Problem 0.1 from §0.
More precisely, one can show (see the proof of Corollary 7.12 below) that a uniformly bounded representation $\pi\colon G \to B(H)$ extends canonically to a bounded unital homomorphism $u_\pi\colon C^*(G) \to B(H)$ if all the coefficients of π are in $B(G)$. This shows that if Problem 0.2 happens to have a positive solution for $A = C^*(G)$, then the inclusion $\bigcup_{c>1} B_c(G) \subset B(G)$ holds (if and) only if every uniformly bounded representation π on G is unitarizable.

Theorem 2.1 comes from [BF2] (preceded by [Fe1]). The idea of introducing the space $T_p(G)$ comes from [W1] where Theorem 2.5 appears and Lemmas 2.2 and 2.7 (ii) are also mentioned there. Lemma 2.7 (i) has a long history, it goes back to Leinert [L1], then the constant was improved by Bożejko, and very precise estimates were obtained in [AO]. Finally, Haagerup [H2] obtained a far reaching generalization for the functions supported by the words of any fixed given length. See [HP1] for more recent information along the same line.

A general reference for Fourier analysis on free groups is the book [FTP] by Figà-Talamanca and Picardello. For the closely related case of trees, see [FTN].

Corollary 2.3 was first obtained in [MZ1-2]. As mentioned earlier, the first examples (on $SL_2(\mathbb{R})$) of uniformly bounded non unitarizable group representations appeared in [EM] and later in [KS]. See Cowling's papers [Cow1-4] for more on this theme. More recently, there has been numerous contributions aiming at constructing, as simply as possible, non trivial analytic families of uniformly bounded representations (these yield non unitarizable ones). For such constructions we refer to [PyS, B5, W2]. For the viewpoint of groups acting on trees, see also [Va1-2].

Concerning power bounded operators, we chose to follow in the text the approach of [B2] in order to emphasize the parallel aspects between the free group case and the power bounded (=semi-group) case. But we should mention that this approach is actually very similar to some ideas appearing in [Pe1], related to Davie's earlier work [Dav]. Note also that the original example of Foguel [Fo, Le] (power bounded but not similar to a contraction) already uses a lacunary sequence. See also [FW].

Our treatment emphasizes the analogy existing between the generators of the free group (or free sets in general) and "lacunary" or "thin" subsets of groups, such as Sidon sets. These sets, although they are not free, satisfy sufficiently few relations to behave in many ways like free sets. General references on this subject are [LP, LR, R1, Ka]. See [P3] for a more recent illustration of this analogy.

Statements 2.9, 2.12, 2.13 all come from [B2], Lemma 2.10 from [Fou].

We refer the reader to the stimulating and fascinating problem book [HaN] for more questions related to similarity problems and Hankel or Toeplitz operators. Beware however, that several problems there, for instance 7.2 in [HaN], have not been properly updated (question 1 there was answered in [LuP], and question 5 in [Bo]). We will give detailed information on that particular set of problems of Peller in the notes and remarks on chapter 6.

3. Completely bounded maps

Summary: In this chapter, we first prove a fundamental criterion for an operator between Banach spaces to factor through a Hilbert space. Then we turn to the notion of complete boundedness (which is crucial for these notes). We prove a fundamental factorization/extension theorem for completely bounded maps, and give several consequences. In this viewpoint, the underlying idea is the same in both cases (completely bounded maps or operators factoring through Hilbert space). At the end of this chapter, we give several examples of bounded linear maps which are not completely bounded, and related norm estimates.

In this chapter we will prove the basic factorization theorem of completely bounded maps. As many factorization theorems in Functional Analysis it can be derived rather directly from the Hahn-Banach theorem.

Theorem 3.1. (Hahn-Banach) *Let Λ be a real vector space. Let $p\colon \Lambda \to \mathbb{R}$ be a sublinear map, i.e. a map such that*

$$\forall\, x,y \in E \qquad p(x+y) \le p(x) + p(y)$$
$$\forall\, t \ge 0 \qquad\qquad p(tx) = tp(x).$$

Then there is an $\mathbb{R}$-linear functional $f\colon \Lambda \to \mathbb{R}$ such that

$$\forall x \in E \qquad f(x) \le p(x).$$

Proof: See e.g. [R3, p. 56]. (Note that since $p(0) = 0$ there is a linear functional which is majorized by p on the subspace $\{0\}$!) $\qquad\qquad\qquad\qquad\qquad\square$

Corollary 3.2. *Let Λ_+ be a convex cone in a real vector space Λ. Let $q\colon \Lambda_+ \to \mathbb{R}$ be a superlinear map i.e. a map such that for all $t \ge 0$, $q(tx) = tq(x)$ for all x in Λ_+ and*

$$\forall\, x,y \in \Lambda_+ \qquad q(x) + q(y) \le q(x+y).$$

Let $p\colon \Lambda \to \mathbb{R}$ be a sublinear map. If $q(x) \le p(x)$ for all x in Λ_+ then there is an $\mathbb{R}$-linear functional $f\colon \Lambda \to \mathbb{R}$ such that

$$\forall\, x \in \Lambda_+ \qquad q(x) \le f(x) \le p(x)$$

(and actually $f(x) \le p(x)$ for all x in Λ).

Proof: Let $r(x) = \inf\{p(x+y) - q(y) \mid y \in \Lambda_+\}$. It is easy to check that r is sublinear, that $-p(-x) \le r(x) \le p(x)$ (so that $r(x)$ is finite) for all x in Λ

and that $r(-y) \leq -q(y)$ for all y in Λ_+. By the preceding theorem there is a linear form $f: \Lambda \to \mathbb{R}$ such that $f(x) \leq r(x)$ for all x in Λ. A fortiori we have $f(x) \leq p(x)$ for all x in Λ and $f(-y) \leq -q(y)$ for all y in Λ_+. This yields the announced result.   □

Before we come to completely bounded maps, we will prove a classical characterization of operators between Banach spaces which can be factorized through a Hilbert space. The following notation will be convenient. Let X be a Banach space and let (x_j) and (y_i) be finite sequences in X. We will write $(y_i) < (x_j)$ if

$$(3.1) \qquad \forall \xi \in X^* \qquad \sum |\xi(y_i)|^2 \leq \sum |\xi(x_j)|^2.$$

For an $n \times n$ matrix $A = (a_{ij})$ with scalar entries we will denote by $\|A\|_{M_n}$ or $\|(a_{ij})\|_{M_n}$ its norm as an operator on the n-dimensional Hilbert space ℓ_2^n, i.e. we define

$$\|(a_{ij})\|_{M_n} = \sup \left\{ \left(\sum_i \left| \sum_j a_{ij} \alpha_j \right|^2 \right)^{1/2} \mid \sum |\alpha_j|^2 \leq 1 \right\}.$$

Lemma 3.3. *Consider n-tuples $(x_1, \ldots, x_n)$ and $(y_1, \ldots, y_n)$ in X. Then $(y_i) < (x_j)$ iff there is an $n \times n$ matrix (a_{ij}) such that $\|(a_{ij})\|_{M_n} \leq 1$ and $y_i = \sum_j a_{ij} x_j$ for all $i = 1, \ldots, n$.*

Proof: Assume $(y_i) < (x_j)$. Let $E \subset \ell_2^n$ be the linear subspace formed by all the elements of the form $(\xi(x_1), \ldots, \xi(x_n))$ with ξ in X^*. Let $A: E \to \ell_2^n$ be the linear operator defined by $A((\xi(x_j))_{j \leq n}) = (\xi(y_i))_{i \leq n}$. Since $(y_i) < (x_j)$, the operator A is unambiguously defined on E. Composing A with the orthogonal projection onto E we may as well assume A defined on the whole of ℓ_2^n with $\|A\| \leq 1$. Let then (a_{ij}) be the matrix associated to A. We have $(\xi(y_i)) = A((\xi(x_j)))$ hence for all i and all ξ in X^* we have $\xi(y_i) = \sum a_{ij} \xi(x_j) = \xi(\sum a_{ij} x_j)$, so that we conclude that $y_i = \sum a_{ij} x_j$. This proves the "only if" part. The converse is obvious.   □

We can now prove the fundamental criterion for an operator between Banach spaces to factor through a Hilbert space.

Theorem 3.4. *Let X, Y be Banach spaces. Let $u: X \to Y$ be an operator and let $C \geq 0$ be a constant. The following assertions are equivalent.*

(i) *There is a Hilbert space H and operators $B: X \to H$ and $A: H \to Y$ such that $u = AB$ and $\|A\| \cdot \|B\| \leq C$.*

(ii) *For all n and all $n \times n$ matrices (a_{ij}) with $\|(a_{ij})\|_{M_n} \leq 1$ we have*

$$\forall \, x_j \in X \qquad \sum_i \left\| \sum_j a_{ij} u(x_j) \right\|^2 \leq C^2 \sum \|x_j\|^2.$$

(iii) *For all finite sets (y_i) and (x_j) in X the relation $(y_i) < (x_j)$ implies*

$$\sum \|u(y_i)\|^2 \le C^2 \sum \|x_j\|^2.$$

Proof: (i) $\Rightarrow$ (ii) is easy. Observe that if $x_j = (x_j(s))_{s \in S}$ are elements in the Hilbert space $H = \ell_2(S)$ where S is an arbitrary set then we can write

$$\sum_i \left\| \sum_j a_{ij} x_j \right\|_H^2 = \sum_i \sum_s \left| \sum_j a_{ij} x_j(s) \right|^2$$

$$\le \|(a_{ij})\|_{M_n}^2 \sum_s \sum_j |x_j(s)|^2 = \sum \|x_j\|_H^2.$$

This shows that (ii) holds with $C = 1$ when u is the identity on H. From this (using $\|A\| \|B\| \le C$) is easy to show that (i) implies (ii). We leave the details to the reader. The implication (ii) $\Rightarrow$ (iii) follows directly from Lemma 3.3 It remains to check (iii) $\Rightarrow$ (i). Assume (iii). We will use Corollary 3.2. Let $I = X^*$. For any x in X we will denote by $\hat{x} \colon I \to \mathbb{R}$ or $\mathbb{C}$ the function defined by

$$\forall \, \xi \in I \qquad \hat{x}(\xi) = \xi(x).$$

We denote by Λ the real vector space formed of all the functions $\varphi \colon I \to \mathbb{R}$ such that there is a finite set (x_j) in X such that $|\varphi| \le \sum |\hat{x}_j|^2$. Let $\Lambda_+ = \{\varphi \in \Lambda \mid \varphi \ge 0\}$. Then we define

$$\forall \varphi \in \Lambda \qquad p(\varphi) = \inf \left\{ C^2 \sum \|x_j\|^2 \mid x_j \in X, \varphi \le \sum |\hat{x}_j|^2 \right\}$$

and

$$\forall \varphi \in \Lambda_+ \qquad q(\varphi) = \sup \left\{ \sum \|u(y_i)\|^2 \mid y_i \in X, \sum |\hat{y}_i|^2 \le \varphi \right\}.$$

Note that the sets relative to which the inf and the sup run are both nonvoid so that $p(\varphi)$ and $q(\varphi)$ are well defined. Moreover clearly p is sublinear and q superlinear so that Corollary 3.2 ensures that there is a linear functional $f \colon \Lambda \to \mathbb{R}$ such that $q(\varphi) \le f(\varphi) \le p(\varphi)$ for all φ in Λ_+. In particular taking $\varphi = |\hat{x}|^2$ with x arbitrary in X we find

$$\forall \, x \in X \qquad \|u(x)\|^2 \le f(|\hat{x}|^2) \le C^2 \|x\|^2.$$

It is now routine to complete the proof: we extend f by linearity to a complex linear form on $\Lambda + i\Lambda$ and we define $\langle x, y \rangle = f(\hat{x}\bar{\hat{y}})$ for all x, y in X, this defines a prehilbertian scalar product on X such that $\|u(x)\|^2 \le \langle x, x \rangle \le C^2 \|x\|^2$ for all x in X; after passing to the quotient by $\{x \mid \langle x, x \rangle = 0\}$ and completing we obtain a Hilbert space H and a mapping $B \colon X \to H$ such that $\|u(x)\|^2 \le \|B(x)\|^2 \le C^2 \|x\|^2$ for all x in X, we may as well assume $\overline{B(X)} = H$, then we can define A by setting $A(B(x)) = u(x)$ and we obtain $\|A\| \le 1$. $\qquad\square$

It is customary to write

$$(3.2) \qquad\qquad \gamma_2(u) = \inf\{\|A\| \|B\|\}$$

where the infimum runs over all possible factorizations of u as in Theorem 3.4 (i) and to denote by $\Gamma_2(X, Y)$ the set of all operators $u \colon X \to Y$ satisfying Theorem 3.4 for some $C > 0$. Among the consequences of Theorem 3.4 (which can also be checked by a direct argument) we have

Corollary 3.5. *The functional $u \to \gamma_2(u)$ is a norm on $\Gamma_2(X, Y)$ with which $\Gamma_2(X, Y)$ becomes a Banach space. Moreover the infimum is attained in (3.2).*

Let H, K be Hilbert spaces. Let $S \subset B(H)$ be a subspace. In the theory of operator algebras, the notion of complete boundedness for a map $u : S \to B(K)$ has been extensively studied recently, cf. [Pa1]. Its origin lies in the work of Stinespring [Sti] and Arveson [Ar1] on completely positive maps (see [Pa1] for more details and references). Let us equip $M_n(S)$ and $M_n(B(K))$ (the spaces of matrices with entries respectively in S and $B(K)$) with the norm induced respectively by $B(\ell_2^n(H))$ and $B(\ell_2^n(K))$.
A map $u : S \to B(K)$ is usually called completely bounded (in short c.b.) if there is a constant C such that the maps $I_{M_n} \otimes u$ are uniformly bounded by C i.e. if we have

$$\sup_n \|I_{M_n} \otimes u\|_{M_n(S) \to M_n(B(K))} \le C,$$

and the c.b. norm $\|u\|_{cb}$ is defined as the smallest constant C for which this holds.
When $\|u\|_{cb} \le 1$, we say that u is completely contractive (or a complete contraction). Also u is called completely isometric if the maps $I_{M_n} \otimes u$ are isometric for all n.
Moreover, u is called completely positive if the maps $I_{M_n} \otimes u$ are positive maps (i.e. they map positive elements to positive ones) for all n.
It is easy to check that for any c.b. map $v : u(S) \to B(L)$ (with L another Hilbert space) the composition vu is c.b. and satisfies

$$\|vu\|_{cb} \le \|v\|_{cb}\|u\|_{cb}.$$

It is quite straightforward to extend the usual definitions to the Banach space case as follows. Let X, Y be Banach spaces. We denote by $X \otimes Y$ their algebraic tensor product and by $B(X, Y)$ the space of all bounded operators from X into Y, equipped with the usual operator norm. Let X_1, Y_1 be an other couple of Banach spaces. Let $S \subset B(X_1, Y_1)$ be a subspace and let $u : S \to B(X, Y)$ be a linear map. For any $n \ge 1$ we denote by $M_n(S)$ the space of all $n \times n$ matrices (a_{ij}) with coefficients in S with the norm

$$\|(a_{ij})\|_{M_n(S)} = \sup \left(\sum_i \left\| \sum_j a_{ij} x_j \right\|^2 \right)^{\frac{1}{2}}$$

where the supremum runs over all $x_1, \ldots, x_n$ in X_1 such that $\sum \|x_j\|^2 \le 1$. We define $u_n : M_n(S) \to M_n(B(X, Y))$ by $u_n((a_{ij})) = (u(a_{ij}))$. We will say that u is completely bounded (in short c.b.) if the maps u_n are uniformly bounded and we define

$$\|u\|_{cb} = \sup_{n \ge 1} \|u_n\|.$$

The notion of complete positivity can also be extended provided we assume that $Y = \overline{X^*}$ and $Y_1 = \overline{X_1^*}$ (this denotes the *antilinear* dual spaces). Then positivity is defined in $B(X, \overline{X^*})$ by the positivity of the corresponding quadratic forms. Since $\overline{\ell_2^n(X)^*} = \ell_2^n(\overline{X^*})$, this notion of positivity makes sense also in $M_n(B(X, \overline{X^*}))$ and a fortiori in S. We can then define a completely positive (in short c.p.) map u as one such that u_n is positive for all n. Clearly these definitions generalize the usual ones when $X = Y = H$ and $X_1 = Y_1 = K$, H and K being Hilbert spaces.

We will prove the fundamental factorization of cb maps in a very similar way to the above proof of theorem 3.4, by a rather direct application of the Hahn-Banach Theorem. This result was proved by several mathematicians independently. It seems that Wittstock [Wi1-2] was the first to prove it, but Haagerup had an unpublished proof (see [H4]) and Paulsen also discovered it but slightly later.

Theorem 3.6. *Let H be a Hilbert space and let $S \subset B(H)$ be a subspace. Let X, Y be Banach spaces. Let $u : S \to B(X, Y)$ be a c.b. map. Then there is a Hilbert space $\widehat{H}$, a (C^*-algebraic) representation $\pi : B(H) \to B(\widehat{H})$ with $\pi(1) = 1$ and operators $V_1 : X \to \widehat{H}$ and $V_2 : \widehat{H} \to Y$ with $\|V_1\| \, \|V_2\| \leq \|u\|_{cb}$ such that*

$$(3.3) \qquad\qquad \forall a \in S \quad u(a) = V_2 \pi(a) V_1.$$

Conversely, any map of the form (3.3) satisfies

$$\|u\|_{cb} \leq \|V_2\| \, \|V_1\|.$$

For the proof we will use the following notation. Let I be the space $B(X, H)$. We denote by $\widehat{u} : S \otimes X \to Y$ the linear map induced by u, i.e., for any $z = \sum a_i \otimes x_i$ in $S \otimes X$ we define

$$(3.4) \qquad\qquad \widehat{u}(z) = \sum u(a_i) x_i.$$

For any ξ in I, we denote

$$\xi . z = \sum a_i \xi(x_i).$$

(We will often write $\xi . x_i$ instead of $\xi(x_i)$.)

Let $(z_i)_{i \leq n}$ be a finite sequence in $S \otimes X$ and let $(x_i)_{i \leq n}$ be a finite sequence in X. We write $(z_i) < (x_i)$ if

$$(3.5) \qquad\qquad \forall \xi \in I \quad \sum \|\xi . z_i\|_H^2 \leq \sum \|\xi . x_i\|_H^2.$$

We will use the following elementary (linear algebraic) fact

Remark. Consider $x_1, \ldots, x_n$ in X. Assume that $z \in S \otimes X$ is such that if $\xi \in I, \xi(x_i) = 0$ for all $i \leq n$ implies $\xi . z = 0$. Then we claim that there are a_j in S such that $z = \sum a_j \otimes x_j$. This is a simple fact. Indeed, consider η in X^* such that $\eta \in \{x_1, \ldots, x_n\}^{\perp}$ and consider y arbitrary in H. Let $\xi = \eta \otimes y : X \to H$. Then by assumption we must have $\xi . z = 0$. Without loss of generality, we can assume that (x_j) are linearly independent. Let x_j^* be the functionals biorthogonal to (x_j) in X^*. Let $a_j = (I_S \otimes x_j^*)(z)$. We claim that $z = \sum a_j \otimes x_j$ because

$z' = z - \sum a_j \otimes x_j$ vanishes on any element of $H \otimes X^*$. Indeed if η' in X^* and y in H are arbitrary, then $\eta = \eta' - \sum \eta'(x_i)x_i^* \in \{x_1, \ldots, x_n\}^\perp$ hence by the first part of this remark if $\xi = \eta \otimes y$ we must have $\xi.z = 0$ or equivalently $< \eta' \otimes y, z' >= 0$ so that we conclude $z' = 0$.

From this remark we deduce another easy fact, as follows

Lemma 3.7. *Let $(z_i)_{i \leq n}$ and $(x_i)_{i \leq n}$ be as above. Then (3.5) holds iff there is a matrix (a_{ij}) in $M_n(S)$ with $\|(a_{ij})\|_{M_n(S)} \leq 1$ such that*

$$\forall i = 1, 2, \ldots, n \qquad z_i = \sum_j a_{ij} \otimes x_j.$$

Proof: Assume (3.5). Then necessarily, by the preceding remark there are a_{ij} in S such that $z_i = \sum_j a_{ij} \otimes x_j$. By (3.5) we have

$$(3.6) \qquad \forall \xi \in I \quad \sum_i \left\| \sum_j a_{ij} \xi(x_j) \right\|^2 \leq \sum_j \|\xi(x_j)\|^2.$$

Let $A = (a_{ij})$. Assume first (x_j) linearly independent, then $(\xi(x_j))_{j \leq n}$ runs over all of H^n when ξ runs over I, therefore (3.6) implies $\|(a_{ij})\|_{M_n(S)} \leq 1$ and we are done.

Now assume that the span of (x_j) has dimension $m < n$. Let $E \subset \ell_2^n$ be the subspace formed of all $(x^*(x_1), \ldots, x^*(x_n))$ when x^* runs over X^*. Let $P : \ell_2^n \to \ell_2^n$ be the orthogonal projection onto E. Furthermore, let $\tilde{E} \subset \ell_2^n(H)$ be the subspace formed of all $(\xi(x_1), \ldots, \xi(x_n))$ when ξ runs over I and let $\tilde{P} : \ell_2^n(H) \to \tilde{E}$ be the orthogonal projection. Clearly we have $\tilde{E} = E \otimes H$ and $\tilde{P} = P \otimes I_H$. By (3.6) we clearly have $\sum_i \| \sum_j a_{ij} y_j \|^2 \leq \sum \|y_j\|^2$ for all $(y_1, \ldots, y_n)$ in $\tilde{E}$. Now let $h = (h_1, \ldots, h_n)$ be arbitrary in $\ell_2^n(H)$ and let $y = (y_1, \ldots, y_n) = \tilde{P}h$ (*i.e.* $y_j = \sum_k P_{jk} h_k$). We can write

$$\sum_i \| \sum_k (AP)_{ik} h_k \|^2 = \sum_i \| \sum_j a_{ij} y_j \|^2 \leq \sum \|y_j\|^2 = \|\tilde{P}h\|^2 \leq \|h\|^2.$$

Hence we have $\|AP\|_{M_n(S)} \leq 1$.
On the other hand, by definition of P we have

$$x_j = \sum P_{jk} x_k \text{ for all } j$$

hence

$$z_i = \sum_j a_{ij} \otimes x_j = \sum_k (AP)_{ik} \otimes x_k.$$

Hence we conclude by replacing A by AP. This shows the "only if" part. The "if" part is obvious.   □

Proof of Theorem 3.6 Let $C = \|u\|_{cb}$. By Lemma 3.7, we have for $z_i \in S \otimes X$ and $x_i \in X$

$$(3.7) \qquad (z_i) < (x_i) \Rightarrow \sum \|\hat{u}(z_i)\|^2 \leq C^2 \sum \|x_i\|^2,$$

where $\widehat{u} : S \otimes X \to Y$ is associated to u as explained in (3.4).

Recall $I = B(X, H)$. Let Λ be the space of all functions $\phi : I \to \mathbb{R}$ such that for some finite sequence $x_1, \ldots, x_n$ in X we have

$$\forall \xi \in I \quad |\phi(\xi)| \leq \sum \|\xi(x_i)\|^2.$$

Clearly Λ is a real vector space. Let Λ_+ be the set of all positive functions in Λ. We define

$$\forall \phi \in \Lambda \quad p(\phi) = \inf \left\{ C^2 \sum \|x_i\|^2 \right\}$$

where the infimum runs over all finite sets (x_i) in X such that $\phi(\xi) \leq \sum \|\xi x_i\|^2$ for all ξ in I. Then we define

$$\forall \phi \in \Lambda_+ \quad q(\phi) = \sup \left\{ \sum \|\widehat{u}(z_i)\|^2 \right\}$$

where the supremum runs over all finite sets (z_i) in $S \otimes X$ such that $\sum \|\xi.z_i\|^2 \leq \phi(\xi)$ for all ξ in I. Clearly p is subadditive on Λ, q is superadditive on Λ_+ and both are positively homogeneous. Moreover, we clearly have by (3.7) $q(\phi) \leq p(\phi)$ for all ϕ in Λ_+. Hence by Corollary 3.2 there is a linear form $f : \Lambda \to \mathbb{R}$ such that

(3.8) $\forall \phi \in \Lambda_+ \quad q(\phi) \leq f(\phi)$

(3.9) $\forall \phi \in \Lambda \quad f(\phi) \leq p(\phi)$.

Let us denote by $\Lambda + i\Lambda$ the complexification of Λ. We can extend f by linearity to a $\mathbb{C}$-linear form on $\Lambda + i\Lambda$, which we still denote by f. We now define the space $\mathcal{H}$ consisting of all functions $g : I \to H$ such that the function $\xi \to \|g(\xi)\|_H^2$ lies in Λ. If g' is another element of $\mathcal{H}$, then by Cauchy-Schwarz the function $\xi \to \langle g(\xi), g'(\xi) \rangle$ is in $\Lambda + i\Lambda$ so that we can define

$$\langle g, g' \rangle = f(\langle g(\cdot), g'(\cdot) \rangle).$$

This is a scalar product on $\mathcal{H}$. After passing to the quotient by the kernel of the associated semi-norm and after completing we obtain a Hilbert space $\widehat{H}$.

For x in X, let $\widehat{x} : I \to H$ be defined by $\widehat{x}(\xi) = \xi(x)$. By (3.9) applied to $\phi = \|\widehat{x}\|^2$ we have

$$\langle \widehat{x}, \widehat{x} \rangle \leq C^2 \|x\|^2.$$

Therefore, there is a linear operator $V_1 : X \to \widehat{H}$ with $\|V_1\| \leq C$ such that $V_1(x)$ is the equivalence class of $\widehat{x}$ in $\widehat{H}$. On the other hand, by (3.8) we have

(3.10) $\forall a_i \in S \quad \forall x_i \in X \qquad \left\| \sum u(a_i)x_i \right\|^2 \leq f(\phi)$

where $\phi(\xi) = \| \sum a_i \widehat{x}_i(\xi) \|^2$.

We may clearly define a unit preserving $*$-representation

$$\pi : B(H) \to B(\widehat{H})$$

by setting (up to equivalence classes)

$$\pi(a)g(\xi) = ag(\xi).$$

Then (3.10) yields $\left\| \sum u(a_i)x_i \right\| \leq \left\| \sum \pi(a_i)V_1x_i \right\|$. This allows us to define a linear map

$$V_2 : \overline{\text{span}}(\pi(S)V_1X) \rightarrow Y \text{ with } \|V_2\| \leq 1$$

such that

$$\sum u(a_i)x_i = V_2 \left(\sum \pi(a_i)V_1x_i \right).$$

Finally, we can extend V_2 to an operator $V_2 : \widehat{H} \rightarrow Y$ with norm ≤ 1 and we have the announced result (3.3). The converse is obvious since (as is easy and well known) we have $\|\pi\|_{cb} \leq 1$ for every *-representation π, since $\pi_n = I_{M_n} \otimes \pi$ is a *-representation on the C^*-algebra $M_n(B(H))$ and *-representations have norm 1.   □

Corollary 3.8. *In the same situation as in Theorem 3.6, every c.b. map $u: S \rightarrow B(X,Y)$ extends to a cb map defined on the whole of $B(H)$ with the same cb norm. Moreover, if $Y = \overline{X^*}$, then every cb map $u: B(H) \rightarrow B(X,Y)$ is a linear combination of completely positive maps.*

Proof: Define $\tilde{u}: B(H) \rightarrow B(X,Y)$ by

$$\tilde{u}(a) = V_2\pi(a)V_1.$$

Then $\tilde{u}$ extends u and $\|\tilde{u}\|_{cb} \leq \|V_1\|\|V_2\| \leq \|u\|_{cb}$. This proves the first part. The second one follows simply (as in the Hilbertian case) by polarization from (3.3), once one notices that if $Y = \overline{X^*}$ and if $V_2 = \overline{V_1^*}$, then $\tilde{u}$ is completely positive.   □

Remark 3.9. Note that if $S = B(H)$ and if H is one dimensional then Theorem 3.6 reduces to Theorem 3.4.

More generally, let $u: S \rightarrow B(X,Y)$ be as in Theorem 3.6. Let $v: S \rightarrow B(\widehat{H})$ be the map defined by

$$\forall a \in S \qquad v(a) = \|u\|_{cb}\pi(a).$$

Then there are maps $w_1: X \rightarrow \widehat{H}$, $w_2: \widehat{H} \rightarrow Y$ such that $u(a) = w_2v(a)w_1$, $\|w_1\| \|w_2\| = 1$ and we have

$$\|v\|_{cb} = \|u\|_{cb}.$$

This shows that the generality achieved by letting X, Y be general Banach spaces is a bit disappointing, since the only way a general map u into $B(X,Y)$ can be cb is when it "comes" from a cb map into $B(\widehat{H})$ for some Hilbert space $\widehat{H}$. This explains why in the next corollaries, we will restrict to the case when $X = Y = K$ for some Hilbert space K.

Corollary 3.10. *Let K be a Hilbert space. Note that $M_n = B(\ell_2^n)$. Consider a map $u\colon M_n \to B(K)$ and let $C \geq 0$ be a fixed constant. Then $\|u\|_{cb} \leq C$ iff there is a set I and operators*

$$V_m^1\colon K \to \ell_2^n, \quad V_m^2\colon K \to \ell_2^n, \quad m \in I$$

such that

$$(3.11)' \qquad \left\|\sum_{m \in I} V_m^{1*} V_m^1\right\|^{1/2} \left\|\sum_{m \in I} V_m^{2*} V_m^2\right\|^{1/2} \leq C$$

and

$$(3.11)'' \qquad \forall\, a \in S \qquad u(a) = \sum_{m \in I} V_m^{2*} a V_m^1$$

where the series $\sum\limits_{m \in I}$ are meant in the weak operator topology.

Proof: Changing V_2 to V_2^* in Theorem 3.6, we have that $\|u\|_{cb} \leq C$ iff there are $V_1\colon K \to \widehat{H}$ and $V_2\colon K \to \widehat{H}$, with $\|V_1\|\,\|V_2\| \leq C$, such that

$$(3.12) \qquad \forall\, a \in M_n \qquad u(a) = V_2^* \pi(a) V_1$$

for some $*$-representation $\pi\colon M_n \to B(\widehat{H})$. By a well known (and easy) fact every $*$-representation $\pi\colon M_n \to B(\widehat{H})$ is unitarily equivalent to one of the following form: $\widehat{H} = \mathcal{H} \oplus \cdots \oplus \mathcal{H}$ (n times) or equivalently $\widehat{H} = \ell_2^n(\mathcal{H})$ and $\pi(a) \in B(\ell_2^n(\mathcal{H}))$ is just defined by $\pi(a) = a \otimes I_{\mathcal{H}}$. Therefore we can assume w.l.o.g. that π is indeed of that form and moreover that $\mathcal{H} = \ell_2(I)$ for some set I. Identifying $\ell_2^n(\mathcal{H}) = \ell_2^n(\ell_2(I))$ with $\ell_2(I, \ell_2^n)$ we can write $V_1 = (V_m^1)_{m \in I}$ and $V_2 = (V_m^2)_{m \in I}$ where V_m^1 and V_m^2 are operators from K to ℓ_2^n. Substituting this in the identity (3.12) we obtain (3.11)$''$. Moreover, we have

$$(3.13) \quad \|V_1\|^2 = \sup_{x \in B_K} \sum_{m \in I} \|V_m^1 x\|^2 = \sup_{x \in B_K} \left\langle \sum_{m \in I} V_m^{1*} V_m^1 x, x \right\rangle = \left\|\sum_{m \in I} V_m^{1*} V_m^1\right\|,$$

and similarly for V_2, so that we obtain (3.11)$'$. $\square$

Corollary 3.11. *Let $u\colon M_n \to B(K)$ and C be as in the preceding corollary. Let us denote by (e_{ij}) the canonical basis of M_n and let $x_{ij} = u(e_{ij})$. Then $\|u\|_{cb} \leq C$ iff there is a Hilbert space $\mathcal{H}$ and operators x_j and y_i from K into $\mathcal{H}$ such that*

$$\forall\, i,j \qquad x_{ij} = y_i^* x_j \quad \text{and} \quad \left\|\sum_1^n y_i^* y_i\right\|^{1/2} \left\|\sum_1^n x_j^* x_j\right\|^{1/2} \leq C.$$

Proof: With the same notation as in the preceding proof, we can write $\widehat{H} = \ell_2^n(\mathcal{H})$ and we can identify each of $V_1\colon K \to \ell_2^n(\mathcal{H})$ and $V_2\colon K \to \ell_2^n(\mathcal{H})$ with an n-tuple of operators from K into $\mathcal{H}$ denoted respectively by $(x_1, \ldots, x_n)$ and $(y_1, \ldots, y_n)$. Then (3.12) becomes

$$\forall\, a = (a_{ij}) \in M_n \qquad u(a) = \sum_{ij} a_{ij} u(e_{ij}) = \sum_{ij} a_{ij} y_i^* x_j$$

hence $x_{ij} = y_i^* x_j$ and by the same reasoning as above for (3.13) we have $\|V_1\| = \left\|\sum x_i^* x_i\right\|^{1/2}$ and $\|V_2\| = \left\|\sum y_i^* y_i\right\|^{1/2}$. $\qquad\square$

Corollary 3.12. For any map $u \colon M_n \to M_n$, we have $\|u\|_{cb} \leq C$ iff there are finite sets of operators $(a_m)_{m \in I}$ and $(b_m)_{m \in I}$ with $\mathrm{card}(I) \leq n^2$ such that

$$\forall\, a \in M_n \qquad u(a) = \sum_{m \in I} a_m a b_m$$

and $\left\|\sum_I a_m a_m^*\right\|^{1/2} \left\|\sum_I b_m^* b_m\right\|^{1/2} \leq C$. (In particular, for $C = \|u\|_{cb}$ we have equality.)

Proof: We use the preceding corollary. Note that if $\dim K = n$ then the span of $\bigcup_{1 \leq j \leq n} x_j(K)$ is of dimension at most n^2. Let us denote it by $\mathcal{H}_0 \subset \mathcal{H}$ and let P be the orthogonal projection from $\mathcal{H}$ onto $\mathcal{H}_0$. Clearly $y_i^* x_j = (Py_i)^*(Px_j)$ for all i and j, so that we can replace $\mathcal{H}$ by $\mathcal{H}_0$ if we wish, but then the set I appearing in Corollary 3.10 is finite with $\mathrm{card}(I) \leq n^2$. Then with the notation of Corollary 3.10, to conclude it suffices to let $a_m = V_m^{2*}$ and $b_m = V_m^1$. $\qquad\square$

The following result due to R. Smith [Sm] is often useful.

Proposition 3.13. Consider $S \subset B(H)$ as above and $u : S \to M_n = B(\ell_2^n, \ell_2^n)$. Then we have

$$\|u\|_{cb} = \|I_{M_n} \otimes u\|_{M_n(S) \to M_n(M_n)}.$$

Proof: This can be proved using the fact that if $x_1, \ldots, x_m$ is a finite subset of ℓ_2^n with $\sum_1^m \|x_i\|^2 \leq 1$ then (we leave this as an exercise for the reader) there is an $m \times n$ scalar matrix $b = (b_{jk})$ with $\|(b_{jk})\| \leq 1$ and vectors $\tilde{x}_1, \ldots, \tilde{x}_n$ in ℓ_2^n such that $\sum_1^n \|\tilde{x}_i\|^2 \leq 1$ and

$$\forall j \leq m \quad x_j = \sum_{k=1}^n b_{jk} \tilde{x}_k.$$

Similarly for any $y_1, \ldots, y_m$ in ℓ_2^n there is a scalar matrix $c = (c_{il})$ with $\|(c_{il})\| \leq 1$ and there are $\tilde{y}_1, \ldots, \tilde{y}_n$ in ℓ_2^n such that $\sum_1^n \|\tilde{y}_i\|^2 \leq 1$ and

$$\forall i \leq m \quad y_i = \sum_{l=1}^n c_{il} \tilde{y}_l.$$

Hence for any $m \times m$ marix (a_{ij}) in $M_m(S)$ we have

$$\sum_{i,j=1}^m <u(a_{ij})x_j, y_i> = <\sum_{k,l=1}^n u(\alpha_{lk})\tilde{x}_k, \tilde{y}_l>$$

where $(\alpha_{lk}) = c^*.(a_{ij}).b$ (matrix product). Therefore we obtain

$$\|(u(a_{ij}))\|_{M_m(M_n)} \leq \|(u(\alpha_{kl}))\|_{M_n(M_n)} \leq \|I_{M_n} \otimes u\|_{M_n(S) \to M_n(M_n)}.$$

$\square$

Corollary 3.14. *Let H, K be Hilbert spaces with K finite dimensional and let $S \subset B(H)$ be finite dimensional. Then for any $u: S \to B(K)$ and for any $\varepsilon > 0$ there is a finite set of operators $(a_m)_{m \in I}$ and $(b_m)_{m \in I}$ from K into H such that*

$$\forall\, a \in S \qquad u(a) = \sum_{m \in I} a_m^* a b_m$$

and

$$\left\|\sum a_m^* a_m\right\|^{1/2} \left\|\sum b_m^* b_m\right\|^{1/2} \leq \|u\|_{cb} + \varepsilon.$$

Proof: Actually this can be deduced from the preceding two statements, by replacing S by a suitably "approximating" subspace of $B(\ell_2^N)$ for some N large enough. We leave the details as an exercise to the reader. $\square$

Finally, we wish to record here the following simple fact.

Proposition 3.15. *Let A be a C^*-algebra. Let H_1, H_2 be two Hilbert spaces and let $J_1: A \to B(H_1)$ and $J_2: A \to B(H_2)$ be two (isometric) $*$-representations of A as a sub-C^*-algebra of $B(H_1)$ and $B(H_2)$ respectively. Then the norms induced on $M_n(A)$ by $M_n(B(H_1))$ and $M_n(B(H_2))$ coincide.*

Proof: The norms induced on $M_n(A)$ by $M_n(B(H_1))$ and $M_n(B(H_2))$ induce a C^*-algebra structure on $M_n(A)$. So the identity map on $M_n(A)$ can be viewed as a $*$-representation from one C^*-algebra structure into the other. Since injective $*$-representations are necessarily isometric, the above statement is immediate.

$\square$

This result allows us to speak of $M_n(A)$ independently of the realization of A as a subalgebra of $B(H)$. In particular, we have

Corollary 3.16. *Let A be a commutative C^*-algebra with unit say $A = C(T)$ for some compact set T. Then for (a_{ij}) in $M_n(A)$ we have*

$$(3.14) \qquad \|(a_{ij})\|_{M_n(A)} = \sup_{t \in T} \|(a_{ij}(t))\|_{M_n}.$$

Proof: The embedding $C(T) \to B(\ell_2(T), \ell_2(T))$ which maps $a \in C(T)$ to the diagonal multiplication operator on $\ell_2(T)$ with diagonal coefficients $(a(t))_{t \in T}$ is clearly a $*$-representation. The norm induced by $B(\ell_2(T), \ell_2(T))$ is clearly given by (3.14). $\square$

Remark 3.17. Recall that when E, F are Banach spaces, their injective tensor product $E \check{\otimes} F$ is the completion of $E \otimes F$ for the injective norm defined as

$$\forall\, u \in E \otimes F \qquad \|u\|_v = \sup\{|\langle u, \xi \otimes \eta \rangle| \mid \xi \in E^*,\ \eta \in F^*,\ \|\xi\| \leq 1,\ \|\eta\| \leq 1\}.$$

In particular, if $E = C(T)$, it is easy to check that $E \check{\otimes} F$ can be identified isometrically with the space $C(T, F)$ of F valued continuous functions on T

equipped with the sup norm. Thus we can rewrite (3.14) as follows: if A is a commutative C^*-algebra we have for all (a_{ij}) in $M_n(A)$

$$(3.15) \qquad \|(a_{ij})\|_{M_n(A)} = \|(a_{ij})\|_{M_n \check{\otimes} A}.$$

Indeed, we can always add a unit if A does not have one.
More generally, for any subspace $S \subset B(H)$ and any (a_{ij}) in $M_n(S)$ we clearly have

$$(3.16) \qquad \|(a_{ij})\|_{M_n \check{\otimes} S} \le \|(a_{ij})\|_{M_n(S)}.$$

Indeed, a simple argument shows that

$$\|(a_{ij})\|_{M_n \check{\otimes} S} = \sup\{\|(\langle a_{ij}h, k\rangle)\|_{M_n} \mid h \in B_H, k \in B_H\} \le \|(a_{ij})\|_{M_n(S)}.$$

As an immediate consequence, we have

Corollary 3.18. *Let $S \subset B(H)$. Consider a bounded operator $u: S \to B(K)$ such that $u(S)$ is included into a commutative C^*-subalgebra A of $B(K)$. Then u is automatically c.b. and $\|u\|_{cb} = \|u\|$.*

Proof: Since the injective tensor product is a tensor norm in the category of Banach spaces, the map $I_{M_n} \otimes u$ has norm $\le \|u\|$ from $M_n \check{\otimes} S$ into $M_n \check{\otimes} A$. But by (3.16) and (3.15) the inclusions $M_n(S) \to M_n \check{\otimes} S$ and $M_n \check{\otimes} A \to M_n(B(K))$ have norm ≤ 1, hence by composition $\|I_{M_n} \otimes u\|_{M_n(S) \to M_n(B(K))} \le \|u\|$, so that taking the supremum over n we obtain $\|u\|_{cb} \le \|u\|$. □

Remark 3.19. It is worthwhile to record here that for any mapping $u: S \to B(H)$ of rank one, we have $\|u\|_{cb} = \|u\|$. Indeed, assume that $u(x) = \xi(x)b$ with $\xi \in S^*$ and $b \in B(H)$. Then u is the composition of the mapping v defined by $v(x) = \xi(x)I$ with the mapping L_b of (say left) multiplication by b. Note that (by Corollary 3.18 for instance) we have $\|v\|_{cb} = \|v\|$. Hence we have

$$\|u\|_{cb} \le \|L_b\|_{cb}\|v\|_{cb} \le \|b\|\,\|v\| = \|b\|\,\|\xi\| = \|u\|.$$

This proves our claim.

It follows that any bounded mapping $u: S \to B(H)$ of finite rank is completely bounded.

Remark 3.20. Let $S \subset A$ be a closed subspace of a C^*-algebra and let $u: S \to B(H)$ be a linear map. We will say that u is c.b. if it is so when we embed A into $B(H)$ via an isometric *-representation. By Proposition 3.15, that is unambiguous and it extends the preceding definition. Moreover, the same remark applies for the cb norm $\|u\|_{cb}$.

There are many examples of bounded maps which are not c.b. For instance let $\tau_n: M_n \to M_n$ be the transposition, i.e. $\tau_n((a_{ij})) = (a_{ji})$. Then clearly $\|\tau_n\| = 1$, but nevertheless

$$(3.17) \qquad \|\tau_n\|_{cb} \ge \sqrt{n}.$$

It follows of course that transposition acting on the space $K(\ell_2)$ (of all compact operators on ℓ_2) is contractive (and actually isometric) but not c.b. To check

(3.17), let $x = (x_{ij}) \in M_n(M_n)$ be the matrix defined by $x_{1j} = e_{j1}$ $(j = 1, \ldots, n)$ and $x_{ij} = 0$ for $i \neq 1$. In other words, x is the $n \times n$ matrix with entries in M_n, which has $(e_{11}, \ldots, e_{n1})$ as its first row and zero elsewhere. Then, a simple calculation yields $\|x\|^2 = \|xx^*\| = 1$ and on the other hand for $y = (\tau_n(x_{ij})) \in M_n(M_n)$, we find $\|y\|^2 = \|yy^*\|^2 \geq n$, whence $\|\tau_n\|_{cb} \geq \sqrt{n}$. Let R_n (resp. C_n) be the subspace of M_n spanned by $\{e_{1j} \mid j = 1, \ldots, n\}$ (resp. $\{e_{i1} \mid i = 1, \ldots, n\}$). Note that R_n and C_n are both isometric to the n-dimensional Hilbert space ℓ_2^n. Let $w_n \colon R_n \to C_n$ be the restriction of τ_n to R_n. The preceding argument shows that $\|w_n\|_{cb} \geq \sqrt{n}$. Actually we have equality, namely $\|w_n\|_{cb} = \sqrt{n}$ and $\|\tau_n\|_{cb} = \sqrt{n}$. We leave this as an exercise for the reader.

Theorem 3.21. *Fix an integer $n \geq 1$. Let $B_2^n \subset \mathbb{C}^n$ be the Euclidean unit ball. Let $\psi_i \colon B_2^n \to \mathbb{C}$ be the restriction of the i-th coordinate. Note that ψ_i belongs to the (commutative) C^*-algebra $C(B_2^n)$ of all continuous complex valued functions on B_2^n. Let*
$$E_2^n = \mathrm{span}[\psi_1, \ldots, \psi_n] \subset C(B_2^n).$$
We consider (implicitly) the C^*-algebra $C(B_2^n)$ as a C^*-subalgebra of $B(H)$ for some H, as in Remark 3.20.

(i) *For each $n \geq 1$, there is a linear map $u_n \colon E_2^n \to B(H)$ with $\|u_n\| \leq 1$ and $\|u_n\|_{cb} \geq n/2$.*

(ii) *Let ℓ_∞^n be the n-dimensional commutative C^*-algebra, i.e. $\mathbb{C}^n$ equipped with the coordinatewise product and the sup norm. Note that ℓ_∞^n can be identified with the subalgebra of all diagonal matrices in M_n. For each $n \geq 1$ there is a linear map $v_n \colon \ell_\infty^n \to B(H)$ with $\|v_n\| \leq 1$ and $\|v_n\|_{cb} \geq \sqrt{n}/2$.*

Proof. Let $x_1, \ldots, x_n$ in $B(H)$ be operators satisfying the canonical anticommutation relations, i.e. such that for all $i, j \leq n$
$$x_i x_j^* + x_j^* x_i = \delta_{ij} I$$
and
$$x_i x_j + x_j x_i = 0.$$
Note that we have

(3.18)
$$\sum_{i=1}^n x_i x_i^* + x_i^* x_i = nI.$$

It is well known that these relations imply

(3.19) $\forall \alpha_i \in \mathbb{C}$ $\left\|\sum \alpha_i x_i\right\| \leq \left(\sum |\alpha_i|^2\right)^{1/2}.$

Indeed, let $T = \sum \alpha_i x_i$. Note that $T^2 = 0$ and $T^*T + TT^* = \sum |\alpha_i|^2 I$. We have $\|T^*TT^*T\| = \|T\|^4$ but $T^*(T^*T)T = 0$, hence
$$\|T\|^4 = \|T(TT^* + T^*T)T\|$$
$$= \left\|T\left(\sum |\alpha_i|^2 I\right) T\right\|$$
$$\leq \|T\|^2 \sum |\alpha_i|^2,$$

which yields (3.19) (note that there is equality when all the coefficients are real). In addition, it is known that these operators can be realized on a finite dimensional Hilbert space H (there is a classical construction using Pauli's matrices, see e.g. [BR] for more or all this).

Let $N = \dim H$ so that $B(H) = M_N$. We will identify each x_i with the matrix $\{x_i(k, \ell) \mid 1 \leq k, \ell \leq N\}$. We now introduce the linear map $u_n \colon E_2^n \to B(H)$ defined by

$$u_n(\psi_i) = x_i.$$

Let a (resp. b) be the element of $M_N(M_N)$ (resp. $M_N(E_2^n)$) defined as follows

$$a_{k\ell} = \sum_{i=1}^{n} \overline{x_i(k, \ell)} x_i$$

$$b_{k\ell} = \sum_{i=1}^{n} \overline{x_i(k, \ell)} \psi_i.$$

Note that $(a_{k\ell}) = (u_n(b_{k\ell}))$ so that we have by definition of $\|u_n\|_{cb}$

(3.20) $$\|a\|_{M_N(M_N)} \leq \|u_n\|_{cb} \|b\|_{M_N(E_2^n)}.$$

Now we view a as acting on $\ell_2^{N^2}$. We compute $\langle ak, k \rangle$, with $k \in \ell_2^{N^2}$ defined by $k(\ell, \ell') = N^{-1/2}$ if $\ell = \ell'$ and $k(\ell, \ell') = 0$ otherwise (note that k has norm 1 in $\ell_2^{N^2}$), we find

$$\|a\|_{M_N(M_N)} \geq \frac{1}{N} \sum_{i=1}^{n} \sum_{k, \ell \leq N} |x_i(k, \ell)|^2$$

$$= \frac{1}{N} \sum_{i=1}^{n} \frac{1}{2} (tr\ x_i^* x_i + tr\ x_i x_i^*).$$

By (3.18) the last expression is equal to $\frac{n}{2}$ hence we obtain $\|a\|_{M_N(M_N)} \geq n/2$. On the other hand by Corollary 3.18, we have

$$\|b\|_{M_N(E_2^n)} = \sup \left\{ \left\| \left(\sum_{i=1}^{n} \overline{x_i(k, \ell)} \alpha_i \right)_{k, \ell} \right\|_{M_N} \ \middle|\ (\alpha_i)_{i \leq n} \in B_2^n \right\}$$

hence by (3.19)

$$\|b\|_{M_N(E_2^n)} \leq 1.$$

Combining these estimates into (3.20) we finally obtain $\|u_n\|_{cb} \geq n/2$ as announced.

To prove the second part, let us denote by $j \colon E_2^n \to \ell_\infty^n$ the natural inclusion map. Let $v_n = n^{-1/2} u_n j^{-1}$ so that $u_n = n^{1/2} v_n j$. Note that $\|j^{-1}\| = n^{1/2}$ so that $\|v_n\| \leq 1$. Also, by Corollary 3.18 we have $\|j\|_{cb} = \|j\| = 1$, therefore

$$n/2 \leq \|u_n\|_{cb} \leq n^{1/2} \|v_n\|_{cb} \|j\|_{cb} = n^{1/2} \|v_n\|_{cb}$$

which yields as announced

$$\sqrt{n}/2 \leq \|v_n\|_{cb}.$$

□

Remark 3.22. Paulsen [Pa5] introduced the following (possibly infinite) parameter associated to any normed space E. He defined

$$(3.21) \qquad\qquad \alpha(E) = \sup \left\{ \frac{\|w\|_{cb}}{\|w\|} \right\}$$

where the supremum runs over all Hilbert spaces H, K and all nonzero maps $w\colon S \to B(K)$ defined on a subspace $S \subset B(H)$ with S isometric to E. Let T_E be the unit ball of the dual space E^* and let $j\colon E \to C(T_E)$ be the natural isometric embedding. Actually the supremum in (3.21) is attained when we restrict to $S = j(E)$ (and we invoke Remark 3.20). This follows easily from (3.16).

We first discuss $\alpha(E)$ in the finite dimensional case. By Theorem 3.21, we have $\alpha(\ell_2^n) \geq n/2$ and $\alpha(\ell_\infty^n) \geq \sqrt{n}/2$. On the other hand, the operator w_n considered before Theorem 3.21 yields $\alpha(\ell_2^n) \geq \sqrt{n}$, which is better when $n < 4$. More precisely, Paulsen [Pa5] proves that

$$\max \left\{ \frac{n}{2}, \sqrt{n} \right\} \leq \alpha(\ell_2^n) \leq n/\sqrt{2}$$

$$(n/2)^{1/2} \leq \alpha(\ell_\infty^n) \leq \min \left\{ \sqrt{n}, \frac{n}{2} \right\}$$

and he conjectures that $\alpha(\ell_2^n) = n/\sqrt{2}$ and $\alpha(\ell_\infty^n) = (n/2)^{1/2}$.

In addition, we have (see [Pa5] for the proof) for any n-dimensional normed space $\alpha(E) = \alpha(E^*)$.

Remark 3.23. Now let E be an arbitrary infinite dimensional Banach space. We will show that $\alpha(E) = \infty$. We consider again the isometric embedding $j\colon E \to C(T_E)$ and use Remark 3.20. Then there exists a bounded linear map $w\colon E \to B(H)$ which is not c.b. Indeed, let $F \subset E$ be an n-dimensional subspace with $n = \dim F \leq \dim E$. By Fritz John's well known theorem (see e.g. [P1]) there is an isomorphism $v\colon F \to \ell_2^n$ with $\|v^{-1}\| = 1$ such that v admits an extension $\tilde{v}\colon E \to \ell_2^n$ with $\|\tilde{v}\| \leq \sqrt{n}$. We now identify ℓ_2^n with E_2^n and consider the composition $u_n\tilde{v}\colon E \to B(H)$. We have $\|u_n\tilde{v}\| \leq \|u_n\|\|\tilde{v}\| \leq \sqrt{n}$ and on the other hand $u_n = (u_n\tilde{v}_{|F}) \circ v^{-1}$ hence

$$n/2 \leq \|u_n\tilde{v}\|_{cb}\|v^{-1}\|_{cb}$$

but by Corollary 3.18 $\|v^{-1}\|_{cb} = \|v^{-1}\| = 1$ hence we find

$$\frac{\sqrt{n}}{2} \leq \frac{\|u_n\tilde{v}\|_{cb}}{\|u_n\tilde{v}\|}.$$

This shows that the cb norm and the usual norm are not equivalent on the space $B(E, B(H))$. By the closed graph theorem, this guarantees the existence of a bounded map $w\colon E \to B(H)$ which is not c.b..   □

Remark 3.24. By an obvious modification of the preceding argument, one gets for *any* n-dimensional normed space E

$$n^{-1/2}\alpha(\ell_2^n) \leq \alpha(E)$$

hence

$$\sqrt{n}/2 \leq \alpha(E),$$

which shows that $\alpha(E) > 1$ whenever $\dim E \geq 5$.

Notes and Remarks on Chapter 3

The general reference available for completely bounded maps is Paulsen's book [Pa1]. The notion of complete boundedness became important in the early 80's through the independent work of Wittstock [Wi1-2], Haagerup [H4] and Paulsen [Pa2]. These authors discovered Theorem 3.6 (with X, Y Hilbertian) and its corollaries independently (but apparently Wittstock came first). It was inspired by Arveson's earlier work on complete positivity [Ar1], itself motivated by Stinespring's pioneering paper [Sti].

Theorem 3.4 and its corollary essentially come from the paper [LiP], which explained, expanded and amplified the ideas of Grothendieck from [G] (= "the résumé"), who studied $\Gamma_2(X, Y)$ in the tensor product framework. After him, Pietsch [Pie] developed his theory of operator ideals (such as p-absolutely summing operators), and Kwapień [Kw] made a deep investigation of $\Gamma_2(X, Y)$ and more generally $\Gamma_p(X, Y)$ (see the end of chapter 5 for this extension).

The presentation we follow, including the proof of the main result (Theorem 3.6), comes from [P4], where the reader will also find more information on the Banach space versions of complete boundedness.

Proposition 3.13 is due to Roger Smith [Sm] and Corollary 3.14 is probably well known. Statements 3.15 to 3.20 are elementary facts. The estimate (3.17) is well known (we do not know who proved it first). Theorem 3.21 and the subsequent remarks 3.22, 3.23 and 3.24 are all due to Vern Paulsen [Pa5], who relies on the estimates in [H3].

4. Completely bounded homomorphisms and derivations

Summary: In this chapter, we study completely bounded homomorphisms $u: A \to B(H)$ when $A \subset B(\mathcal{H})$ is a subalgebra. We first consider the case when H and $\mathcal{H}$ are Banach spaces but mostly concentrate on the Hilbert space case. In the latter case, we prove the fundamental result that a unital homomorphism is completely bounded iff it is similar to a completely contractive one. Let $\delta: A \to B(H)$ be a derivation on a C^*-algebra. We show that δ is completely bounded iff it is inner. When A is the disc algebra, we prove that an operator T on H is similar to a contraction iff it is completely polynomially bounded, or in other words iff the associated homomorphism $f \to f(T)$ is completely bounded. We discuss a variant for operators on a Banach space and give several related facts. Finally, we give examples showing that a bounded (and actually contractive) unital homomorphism on a uniform algebra is not necessarily completely bounded.

In this chapter we return to the study of compressions of homomorphisms first considered at the end of §1. The notion of "compression" discussed in the Hilbert space setting at the end of §1 can be developed just as well in the Banach space setting. Actually this generalization rather clarifies certain issues.

Let X be a Banach space, and let $T: X \to X$ be a bounded operator. Let

$$E_2 \subset E_1 \subset X$$

be closed subspaces. Assume that E_1 and E_2 are T-invariant i.e. we have $T(E_1) \subset E_1$ and $T(E_2) \subset E_2$. Then T determines a map

$$\tilde{T} \in B(E_1/E_2)$$

which is characterized by the commutative diagram

$$
\begin{array}{ccc}
E_1 & \xrightarrow{\;\;T_{|E_1}\;\;} & E_1 \\
Q \downarrow & & \downarrow Q \\
E_1/E_2 & \xrightarrow{\;\;\tilde{T}\;\;} & E_1/E_2
\end{array}
$$

or equivalently $\tilde{T}Q = QT_{|E_1}$ where $Q: E_1 \to E_1/E_2$ is the canonical surjection onto E_1/E_2. (Indeed, one justifies the existence of $\tilde{T}$ as follows: since E_2 is invariant, $\mathrm{Ker}(QT_{|E_1}) \supset E_2$ hence there is a unique $\tilde{T}$ for which the above diagram commutes.) To describe $\tilde{T}$ more concretely, note that if $x, y \in E_1$ are such that

$$x - y \in E_2 \quad \text{then} \quad Tx - Ty \in E_2,$$

so that $QTx = QTy$. Therefore we can unambiguously *define* $\tilde{T}$ on E_1/E_2 by setting

$$\tilde{T}Q(x) = QTx,$$

and this relation characterizes $\tilde{T}$.

Now let A be a Banach algebra and let

$$u \colon A \to B(X)$$

be a bounded homomorphism, i.e. a bounded linear map such that

$$(4.1) \qquad \forall \, a, b \in A \qquad u(ab) = u(a)u(b).$$

Then, applying the above procedure to $T = u(a)$, we obtain a map

$$\tilde{u}(a) \in B(E_1/E_2)$$

which is characterized by the commutative diagram

$$
\begin{array}{ccc}
E_1 & \xrightarrow{\;u(a)_{|E_1}\;} & E_1 \\[2mm]
Q \downarrow & & \downarrow Q \\[2mm]
E_1/E_2 & \xrightarrow{\;\tilde{u}(a)\;} & E_1/E_2
\end{array}
$$

In particular, this characterization implies

Proposition 4.1. *The map* $\tilde{u} \colon A \to B(E_1/E_2)$ *is a homomorphism with* $\|\tilde{u}\| \leq \|u\|$. *Moreover, if* A *is a subalgebra of* $B(H)$ *(with* H *Hilbert) and if* u *is c.b. then* $\tilde{u}$ *also is c.b. and* $\|\tilde{u}\|_{cb} \leq \|u\|_{cb}$.

Proof: For all a, b in A we have

$$\tilde{u}(a)\tilde{u}(b)Q = \tilde{u}(a)Qu(b) = Qu(a)u(b) = Qu(ab)$$

hence since this relation characterizes $\tilde{u}(ab)$ we must have

$$\tilde{u}(ab) = \tilde{u}(a)\tilde{u}(b).$$

Clearly

$$\|\tilde{u}(a)\|_{B(E_1/E_2)} \leq \|u(a)\|_{B(X)}$$

hence $\|\tilde{u}\| \leq \|u\|$. The fact that $\tilde{u}$ also is c.b. if u is c.b. is easy and left as an exercise for the reader (either go back to the definition of complete boundedness or apply Remark 3.9 to reduce to the case when X is a Hilbert space, then $\tilde{u}$ is unitarily equivalent to the map $a \to P_{E_1 \ominus E_2} u(a)_{|E_1 \ominus E_2}$ whence the conclusion by the trivial direction in Theorem 3.6). $\qquad\qquad\square$

We will say that $\tilde{u}$ is the compression of u to E_1/E_2.

Remark. Note in particular that if $A \subset B(H)$ and if $u: A \to B(Y)$ is the restriction to A of a $*$-representation $\pi: B(H) \to B(Y)$, (with Y Hilbert) then we have

$$\|\tilde{u}\|_{cb} \le \|u\|_{cb} \le \|\pi\|_{cb} = 1.$$

The next result is the Banach space version of Theorem 1.7.

Proposition 4.2. *Let A be a Banach algebra. Let X, Z be two Banach spaces, let $\pi: A \to B(Z)$ be a bounded homomorphism, and let $w_1: X \to Z$ and $w_2: Z \to X$ be operators such that $w_2 w_1 = I_X$. Assume that the mapping $a \to u(a) \in B(X)$ defined by*

$$\forall\, a \in A \qquad u(a) = w_2 \pi(a) w_1$$

is a homomorphism. Then u is similar to a compression of π. More precisely, there are π-invariant subspaces $E_2 \subset E_1 \subset Z$ and an isomorphism $S: X \to E_1/E_2$ such that

$$\|S\|\,\|S^{-1}\| \le \|w_1\|\,\|w_2\|$$

and such that the compression $\tilde{\pi}$ of π to E_1/E_2 satisfies

$$\forall\, a \in A \qquad u(a) = S^{-1}\tilde{\pi}(a)S.$$

Proof: Note that since $w_2 w_1 = I_X$, w_2 is onto. Let

$$E_1 = \overline{\text{span}} \left[w_1(X), \bigcup_{a \in A} \pi(a)w_1(X) \right].$$

Clearly by definition this is a closed π-invariant subspace of Z (the smallest one containing $w_1(X)$). Let

$$E_2 = E_1 \cap \text{Ker}(w_2).$$

We claim that E_2 also is π-invariant. Indeed, consider z in E_1 such that $w_2(z) = 0$. We can write z as a limit of finite sums of the form

$$z = \lim_{n \to \infty} \left[w_1(x^n) + \sum_i \pi(a_i^n)w_1(x_i^n) \right]$$

for some $x^n, x_i^n \in X$, $a_i^n \in A$. Since $w_2 w_1 = I_X$ and $w_2(z) = 0$ we have

$$(4.2) \qquad 0 = \lim_{n \to \infty} \left(x^n + \sum_i u(a_i^n)x_i^n \right).$$

Hence for any a in A

$$\pi(a)z = \lim_{n \to \infty} \left[\pi(a)w_1(x^n) + \sum_i \pi(a)\pi(a_i^n)w_1 x_i^n \right]$$

which implies

$$w_2\pi(a)z = \lim_{n\to\infty}\left[u(a)x^n + \sum_i u(aa_i^n)x_i^n\right]$$

$$= \lim_{n\to\infty} u(a)\left[x^n + \sum_i u(a_i^n)x_i^n\right]$$

(since u is assumed a homomorphism) hence by (4.2)

$$= 0.$$

Since $\pi(a)z \in E_1$, this shows that $\pi(a)z \in E_2$ and proves the above claim. Let $Q\colon E_1 \to E_1/E_2$ be the canonical surjection. Let $S\colon X \to E_1/E_2$ be the operator defined by

$$\forall\, x \in X \qquad S(x) = Qw_1(x).$$

Since $w_2w_1 = I_X$, $w_{2|E_1}\colon E_1 \to X$ is onto and there is a unique isomorphism $R\colon E_1/E_2 \to X$ with $\|R\| \le \|w_2\|$ such that $RQ = w_{2|E_1}$. We have clearly $RQw_1 = w_2w_1 = I_X$ hence $RS = I_X$. This implies that S is invertible and $S = R^{-1}$. Moreover we have

$$\|S\|\,\|S^{-1}\| \le \|Qw_1\|\cdot\|R\| \le \|w_1\|\,\|w_2\|$$

and

$$\forall\, a \in A \qquad S^{-1}\tilde{\pi}(a)S = S^{-1}\tilde{\pi}(a)Qw_1$$
$$= S^{-1}Q\pi(a)w_1$$
$$= RQ\pi(a)w_1$$
$$= w_2\pi(a)w_1$$
$$= u(a).$$

□

Following Paulsen [Pa3-4] (cf. also [H1]) we now apply the preceding remarks to the similarity problems discussed in the introduction.

Theorem 4.3. *Let H, K be Hilbert spaces. Let $A \subset B(H)$ be a subalgebra containing a unit 1 and let $u\colon A \to B(K)$ be a bounded homomorphism with $u(1) = I_K$. Let C be any constant. The following are equivalent:*

(i) *The map u is c.b. with $\|u\|_{cb} \le C$.*
(ii) *There is an isomorphism $S\colon K \to K$ with $\|S\|\,\|S^{-1}\| \le C$ such that the map $a \to S^{-1}u(a)S$ is c.b. with c.b. norm ≤ 1.*

Proof: (ii) $\Rightarrow$ (i) is obvious. Conversely assume (i). By the factorization of c.b. maps proved in the preceding chapter there is a (unit preserving) *-representation

$$\pi\colon B(H) \to B(\widehat{H})$$

and operators $w_1\colon K \to \widehat{H}$ $w_2\colon \widehat{H} \to K$ such that $\|w_1\|\,\|w_2\| \le \|u\|_{cb}$ and

$$u(a) = w_2\pi(a)w_1 \quad \text{for all} \quad a \quad \text{in} \quad A.$$

Since $\pi_{|A}$ and $a \to w_2\pi(a)w_1$ are homomorphisms and $w_2w_1 = u(1) = I_K$, the preceding result implies that u is similar to a compression of $\pi_{|A}$, which we denote by $\tilde{\pi}$. By Proposition 4.1 $\|\tilde{\pi}\|_{cb} \le 1$ and we have $u(a) = S\tilde{\pi}(a)S^{-1}$ with $\|S\|\,\|S^{-1}\| \le \|w_2\|\,\|w_1\| \le \|u\|_{cb}$. This shows that (ii) holds. □

Corollary 4.4. Let $A \subset B(H)$ be a C^*-subalgebra with unit and let $u: A \to B(K)$ be a bounded unital homomorphism. Then u is similar to a *-representation iff u is c.b.. Moreover, we have

$$\|u\|_{cb} = \inf\{\|S\|\|S^{-1}\|\}$$

where the infimum runs over all S such that $a \to S^{-1}u(a)S$ is a *-representation.

Proof: If there is a *-representation $\pi: A \to B(K)$ and an isomorphism $S: K \to K$ such that $u(a) = S^{-1}\pi(a)S$, then u is c.b. and $\|u\|_{cb} \leq \|S\| \|\pi\|_{cb} \|S^{-1}\| \leq \|S\| \|S^{-1}\|$.

To prove the converse recall that on a C^*-algebra, a bounded unital homomorphism $u: A \to B(K)$ is a *-representation iff it is contractive (all unitaries are mapped to invertible isometries, hence to unitaries). Hence if u is c.b. the preceding theorem shows that u is similar to a *-representation. $\qquad \square$

Remark. Let A be a unital C^*-algebra, let $\pi: A \to B(H)$ be a unital *-representation, let $S: H \to H$ be an invertible operator and let $u_S(a) = S^{-1}\pi(a)S$. Then the following assertions are equivalent:

(i) $\|u_S\|_{cb} = 1$,

(ii) u_S is a *-representation,

(iii) There is a unitary U in $B(H)$ such that SU^{-1} belongs to the commutant of $\pi(A)$ in $B(H)$ (denoted by $\pi(A)'$) or equivalently such that $u_S = u_U$.

Indeed, assume (i). Then by Theorem 4.3 there is an invertible isometric ($=$ unitary) operator $S_1: H \to H$ such that $a \to S_1^{-1}u_S(a)S_1$ is a *-representation. Since S_1 is unitary this means equivalently that u_S itself is a *-representation, whence the implication (i) $\Rightarrow$ (ii).

Now assume (ii). Then for any unitary $v \in \pi(A)$ the operator $S^{-1}vS$ must be unitary in $B(H)$ so that $S^{-1}vSS^*v^*S^{*-1} = 1$, or equivalently $vSS^* = SS^*v$, which means that $SS^* \in \pi(A)'$. By the polar decomposition we have $S = hU$ with U unitary in H and with $h = \sqrt{SS^*} \in \pi(A)'$ so that $SU^{-1} \in \pi(A)'$. Thus we have shown that (ii) implies (iii). Finally (iii) $\Rightarrow$ (i) is clear. $\qquad \square$

We now consider the problem whether derivations into $B(H)$ are inner. Consider a C^*-subalgebra $A \subset B(H)$. A map $d: A \to B(H)$ is called a derivation if $d(xy) = xd(y) + d(x)y$ for all x, y in A. It is called inner if there is an operator T in $B(H)$ such that $d(x) = xT - Tx$ for all x in A.

Lemma 4.5. Let A be a C^*-algebra and let $\delta: A \to B(H)$ be a derivation. Then

$$x \longrightarrow u(x) = \begin{pmatrix} x & \delta(x) \\ 0 & x \end{pmatrix}$$

is a homomorphism from A into $B(H \oplus H)$. Moreover, u is similar to a *-representation iff δ is inner.

Proof: If δ is inner, say if $\delta(x) = Tx - xT$ for some T in $B(H)$, then let
$S = \begin{pmatrix} I & T \\ 0 & I \end{pmatrix}$. Clearly S is invertible with inverse $S^{-1} = \begin{pmatrix} I & -T \\ 0 & I \end{pmatrix}$, and we
have $u(x) = S \begin{pmatrix} x & 0 \\ 0 & x \end{pmatrix} S^{-1}$ so that u is similar to a $*$-representation.

Conversely, assume that u is similar to a $*$-representation $\pi: A \to B(H \oplus H)$,
i.e. for some invertible $S: H \oplus H \to H \oplus H$ we have $u(x) = S\pi(x)S^{-1}$. Then let
$a = SS^*$ so that $u(x)a = S\pi(x)S^*$ for all x in A.
Since $\pi(x) = \pi(x^*)^*$, we have $u(x)a = (u(x^*)a)^*$ hence

$$(4.3) \qquad \forall\, x \in A \qquad u(x)a = au(x^*)^*.$$

Let $a = \begin{pmatrix} a_{11} & a_{12} \\ a_{12}^* & a_{22} \end{pmatrix}$. Then (4.3) becomes

$$\begin{pmatrix} x & \delta(x) \\ 0 & x \end{pmatrix} \begin{pmatrix} a_{11} & a_{12} \\ a_{12}^* & a_{22} \end{pmatrix} = \begin{pmatrix} a_{11} & a_{12} \\ a_{12}^* & a_{22} \end{pmatrix} \begin{pmatrix} x & 0 \\ \delta(x^*)^* & x \end{pmatrix}$$

which implies (consider the (2,2) and (1,2) entries only)

$$(4.4) \qquad xa_{22} = a_{22}x \quad \text{and} \quad xa_{12} + \delta(x)a_{22} = a_{12}x.$$

Since $a \geq 0$ is invertible, $\exists\, \delta > 0$ such that $\langle ay, y \rangle \geq \delta \|y\|^2$ for all y in $H \oplus H$
hence the same holds for a_{22} on H and a_{22} must be invertible. Moreover a_{22}
commutes with x so that (4.4) implies

$$\delta(x) = (a_{12}x - xa_{12})a_{22}^{-1} = Tx - xT$$

with $T = a_{12}a_{22}^{-1}$. $\square$

From Corollary 4.4 and the preceding lemma, we immediately deduce

Corollary 4.6. *Let $\delta: A \to B(H)$ be a derivation defined on a unital C^*-algebra
A. Then δ is inner iff it is completely bounded.*

Remark. Let $\pi: A \to B(H)$ be a $*$-homomorphism and let $\delta: A \to B(H)$ be
a derivation relative to π, i.e. such that $\delta(ab) = \pi(a)\delta(b) + \delta(a)\pi(b)$. Then the
preceding result remains valid with the same proof: δ is c.b. iff it is inner (i.e. of
the form $a \to [T, \pi(a)]$ for some T in $B(H)$).

We now return to polynomially bounded operators. An operator $T: H \to H$
will be called completely polynomially bounded (in short c.p.b.) if there is a
constant C such that for all n and all $n \times n$ matrices (P_{ij}) with polynomial
entries we have

$$(4.5) \qquad \|(P_{ij}(T))\|_{B(\ell_2^n(H))} \leq C \sup_{\substack{z \in \mathbb{C} \\ |z|=1}} \|(P_{ij}(z))\|_{B(\ell_n^2)}$$

where $(P_{ij}(T))$ is identified with an operator on $\ell_2^n(H)$ in the natural way. Recall that M_n is identified with $B(\ell_2^n)$. We will denote

$$\|(P_{ij})\|_{M_n(A)} = \sup_{\substack{z \in \mathbb{C} \\ |z|=1}} \|(P_{ij}(z))\|_{B(\ell_n^2)}.$$

Note that T is c.p.b. iff the homomorphism $P \to P(T)$ defines a completely bounded homomorphism u_T from the disc algebra A into $B(H)$. Here of course we consider A as a subalgebra of the C^*-algebra $C(\mathbf{T})$ which itself can be embedded e.g. in $B(L_2(\mathbf{T}))$ by identifying a function f in $C(\mathbf{T})$ or $L_\infty(\mathbf{T})$ with the operator of multiplication by f on $L_2(\mathbf{T})$.

We can now prove Paulsen's theorem.

Corollary 4.7. *An operator T in $B(H)$ is similar to a contraction iff it is completely polynomially bounded. Moreover T is c.p.b. with constant C (as in (4.5) above) iff there is an isomorphism $S\colon H \to H$ such that $\|S\|\,\|S^{-1}\| \le C$ and $\|S^{-1}TS\| \le 1$.*

Proof: Assume that $\widetilde{T} = S^{-1}TS$ is a contraction. Then by Sz.-Nagy's dilation theorem there is a unitary U on a Hilbert space $\widetilde{H} \supset H$ such that U is a strong dilation of $\widetilde{T}$. The map $u_U\colon P \to P(U)$ extends (by the classical functional calculus, see the remark after Theorem 1.1) to a $*$-representation defined on $C(\mathbf{T})$. In particular we have $\|u_U\|_{cb} \le 1$. Since U is a strong dilation of $\widetilde{T}$, we have

$$u_T(P) = P(T) = SP(\widetilde{T})S^{-1} = SP_H P(U)_{|H}S^{-1}$$

hence $u_T(f) = SP_H u_U(f)_{|H}S^{-1}$ for all f in A which implies

$$\|u_T\|_{cb} \le \|S\|\,\|u_U\|_{cb}\,\|S^{-1}\| \le \|S\|\,\|S^{-1}\|.$$

This proves the only if part. The converse is an immediate consequence of Theorem 4.3 applied to the disc algebra A and the homomorphism u_T. □

We should emphasize the close connection between the complete contractivity (or complete positivity) of unital homomorphisms and dilation theorems, as follows.

Theorem 4.8. *Let $A \subset B(H)$ be a unital subalgebra and let $u\colon A \to B(K)$ be a unital homomorphism. Then $\|u\|_{cb} \le 1$ iff there is a Hilbert space $\widehat{H}$, a $*$-representation $\pi\colon B(H) \to B(\widehat{H})$ and an isometric embedding $j\colon K \to \widehat{H}$ such that*

$$\forall a \in A \qquad\qquad u(a) = j^*\pi(a)j.$$

Note that if we view K as sitting inside $\widehat{H}$, then j^* is the orthogonal projection onto K.

Proof: Assume $\|u\|_{cb} \le 1$. Then, by Theorem 3.6 we have $\widehat{H}$, π and V_1, V_2 with $\|V_1\|\,\|V_2\| \le 1$ such that $u(\cdot) = V_2\pi(\cdot)V_1$. Then by Theorem 4.2 (applied to $X = K$ and $\pi_{|A}$) we can find a subspace $E \subset \widehat{H}$ (semi-invariant with respect to $\pi(A)$, i.e. of the form $E = E_1 \ominus E_2$ with E_1, E_2 both $\pi(A)$-invariant) and an isomorphism $S\colon K \to E$ with $\|S\| \, \|S^{-1}\| \leq 1$ such that $u(\cdot) = S^{-1} P_E \pi(\cdot)_{|E} S$. Without loss of generality we can assume $\|S\| \leq 1$ and $\|S^{-1}\| \leq 1$ so that S is unitary. Then it suffices to set $j(x) = S(x)$ (*i.e.* j is the same as S but considered as acting into $\widehat{H}$) to obtain the only if part. The converse is obvious.   □

Remark. Note in passing that $\|u\|_{cb} \leq 1$ implies u completely positive, but this is an instance of a very general fact: any unital complete contraction is completely positive, see [Pa1].

Corollary 4.9. *Let* $\tau = (T_1, \ldots, T_k)$ *be a* k-*tuple of mutually commuting operators on an arbitrary Hilbert space* H. *The following assertions are equivalent*

(i) *The homomorphism taking the polynomial* $P(z_1, \ldots, z_k)$ *to the operator* $P(T_1, \ldots, T_k)$ *extends to a completely contractive unital homomorphism* $u_\tau\colon A(D^k) \to B(H)$.

(ii) *There is a Hilbert space* $\widehat{H}$ *and* k *mutually commuting unitaries* $U_1, \ldots, U_k$ *on* $\widehat{H}$ *such that for any polynomial* P *we have*

$$P(T_1, \ldots, T_k) = P_H P(U_1, \ldots, U_k)_{|H}.$$

Remark. The preceding result applies of course in the particular cases $A = A(D)$ or $A = A(D^2)$. This shows that, in essence, either Sz.-Nagy's or Ando's dilation theorem are equivalent to the complete contractivity of the homomorphism naturally associated with von Neumann's inequality.

Remark. In [SNF], an operator T on a Hilbert space H is said to belong to the class $\mathcal{C}_\rho$ ($\rho \geq 1$) if there is a unitary operator U on a larger Hilbert space $\widetilde{H}$ (containing H as a closed subspace) such that

$$T^n = \rho P_H U_{|H}^n \qquad \forall n \geq 1.$$

Note that this implies for any polynomial P

$$P(T) = P(0)I + \rho P_H (P(U) - P(0)I)_{|H} = P(0)(1 - \rho)I + \rho P_H P(U)_{|H}.$$

Hence $\|P(T)\| \leq (\rho - 1)|P(0)| + \rho \|P\|_\infty \leq (2\rho - 1)\|P\|_\infty$. Similarly, we have for any $n \times n$ matrix (P_{ij}) with polynomial entries

$$\|(P_{ij}(T))\|_{M_n(B(H))} \leq (2\rho - 1) \sup_{z \in D} \|(P_{ij}(z))\|_{M_n},$$

whence

$$\|u_T\|_{cb} \leq 2\rho - 1.$$

Therefore, by Corollary 4.7, every operator in the class $\mathcal{C}_\rho$ ($\rho \geq 1$) is similar to a contraction. This result was originally proved by Sz.-Nagy and Foias (see [SNF, Theorem 8.1]). Note the following curious result due to C.A. Berger: an operator T belongs to the class $\mathcal{C}_\rho$ with $\rho = 2$ iff its numerical radius $w(T) = \sup\{|\langle Th, h\rangle| \mid \|h\| \leq 1\}$ satisfies $w(T) \leq 1$ (see [SNF, Prop. 11.2]).

Remark. In the preceding proof, we observed that by Sz.-Nagy's dilation theorem if an operator T on Hilbert space H satisfies $\|P(T)\| \leq \|P\|_\infty$ for any polynomial P, then necessarily

$$\|(P_{ij}(T))\|_{M_n(B(H))} \leq \|(P_{ij})\|_{M_n(A)}$$

for any $n \times n$ matrix (P_{ij}) with polynomial entries. Similarly, by Ando's theorem, when T_1, T_2 are mutually commuting operators on H the inequality $\|P(T_1, T_2)\| \leq \|P\|_{C(\mathbf{T}^2)}$ for all polynomials in 2 variables implies

$$\|(P_{ij}(T_1, T_2))\|_{M_n(B(H))} \leq \|(P_{ij})\|_{M_n(C(\mathbf{T}^2))}.$$

It is natural to try to find a direct algebraic proof of these two implications. Motivated by this question, Blecher and Paulsen [BP2] have proved that every matrix (P_{ij}) with $\|(P_{ij})\|_{M_n(A)} < 1$ can be written as a finite product of the form

$$P_{ij}(z) = b_1 D_1(z) b_2 D_2(z) \ldots D_N(z) b_{N+1}$$

for some integer N (depending on (P_{ij})) where $b_1, b_2, \ldots, b_{N+1}$ are rectangular matrices independent of z with $\|b_k\| \leq 1$ and where $D_k(z)$ are square diagonal matrices of the form

$$\begin{pmatrix} z^{k_1} & & & & & \\ & \ddots & & & \bigcirc & \\ & & \ddots & & & \\ & & & \ddots & & \\ & \bigcirc & & & \ddots & \\ & & & & & z^{k_m} \end{pmatrix}$$

They also proved a similar result for polynomials in two variables (the diagonal coefficients are then monomials in two variables). Equivalently, any matrix valued polynomial P can be written as a product $P = Q_1 Q_2 \ldots Q_K$ with matrix valued polynomials $Q_1, Q_2, \ldots, Q_K$ each of degree one, $i.e.$ of the form $Q_j(z) = a_j + z b_j$ (it is assumed that the sizes of the rectangular matrices $Q_1, Q_2, \ldots, Q_K$ allow to form the product) and such that $\sup_{z \in D} \|Q_j(z)\| < 1$.

Of course this implies the above implications derived from Sz.-Nagy's or Ando's dilation theorem. Unfortunately however Blecher and Paulsen proved these results *using* the dilation theorems. It would be interesting to find a direct proof of this result of Blecher-Paulsen.

Remark. Note that the disc algebra enjoys a property among nonself-adjoint algebras similar to the property described in Proposition 3.15 for C^*-algebras. Namely if A is isometrically embedded in two different ways as a subalgebra of $B(H)$, then the norms induced on $M_n(A)$ are the same. This is the essential content of Sz.-Nagy's theorem (see the proof of Corollary 4.7). Whether there is an extension of this phenomenon for *isomorphic* embeddings of A as a subalgebra of $B(H)$ is equivalent to Problem 0.3.

We can also deduce from Corollary 4.7 a proof of the following classical result of G.C. Rota [Ro], based on Paulsen's criterion.

Corollary 4.10. *Let* $T\colon H \to H$ *be an operator with spectral radius* $r(T) < 1$. *Then* T *is similar to a contraction.*

Proof: Recall that $r(T) = \varlimsup_{n \to \infty} \|T^k\|^{1/k}$ hence if $r(T) < s < 1$, there is a constant C such that

$$\forall k \geq 0. \qquad\qquad \|T^k\| \leq Cs^k.$$

For any polynomial $f = \sum a_k z^k$ we have $u_T(f) = \sum a_k T^k$. Equivalently, if $\xi_k \in A(D)^*$ is the linear form $f \to a_k$ we have

$$u_T(f) = \sum_{k \geq 0} \xi_k(f)T^k.$$

By Remark 3.19, this implies

$$\|u_T\|_{cb} \leq \sum_0^\infty \|T^k\| \leq C \sum_0^\infty s^k < \infty.$$

Hence by Paulsen's criterion (Corollary 4.7) T is similar to a contraction.   □

Corollary 4.11. *Let* $T\colon H \to H$ *be any bounded operator. Then its spectral radius* $r(T)$ *satisfies*

$$r(T) = \inf\{\|S^{-1}TS\| \mid S\colon H \to H \text{ invertible}\}.$$

Proof: By homogeneity, the preceding corollary shows that the left side of (4.10) is more or equal the right side. Conversely, we have $r(T) = r(S^{-1}TS) \leq \|S^{-1}TS\|$ hence taking the infimum over all S, we obtain the announced equality.   □

The next corollary is a very early result of Sz.-Nagy [SN].

Corollary 4.12. *If* $r(T) \leq 1$ *and if* T *is compact then* T *is similar to a contraction.*

Proof: The idea is that since the non zero part of the spectrum of a compact operator T consists of isolated eigenvalues with finite multiplicity, we can "remove" the part of the spectrum which lies on ∂D in order to obtain an operator T_1 with $r(T_1) < 1$ which only differs from T by a finite rank operator. More precisely by the spectral (analytic) functional calculus we can find a finite rank (nonorthogonal) projection $P\colon H \to H$ such that $PT = TP$ and $E = P(H)$ is invariant for T and $T = TP + T(1-P)$ with $r(T(1-P)) < 1$. Hence $T(1-P)$ is similar to a contraction by the preceding corollary and since TP has finite rank it is easy to conclude that T itself is similar to a contraction. (Observe that if $E = P(H)$ and $F = (I-P)(H)$ then $T\colon H \to H$ is similar to the operator $\widetilde{T}\colon E \oplus F \to E \oplus F$ which maps (x,y) to (Tx, Ty) and if both TP and $T(1-P)$ are similar to contractions, $\widetilde{T}$ also is.)

In the Banach space case, we can state the following (slightly disappointing) result.

Corollary 4.13. *Let* $T\colon X \to X$ *be an operator on a Banach space* X *and let* C *be a constant. The following are equivalent.*

(i) *The map* $u_T\colon P \to P(T)$ *defined on polynomials extends to a c.b. map from* $A(D)$ *into* $B(X)$ *with* $\|u_T\|_{cb} \leq C$.

(ii) *There is a Hilbert space* H *and an isomorphism* $S\colon X \to H$ *such that*

$$\|STS^{-1}\| \leq 1 \quad \text{and} \quad \|S\|\|S^{-1}\| \leq C.$$

Proof: Assume (i). By Theorem 3.6 there are $\hat{H}$, V_1, V_2 with $\|V_1\|\|V_2\| \leq C$ and such that $u_T(P) = V_2\pi(P)V_1$ for any polynomial P. In particular since

$$(4.6) \qquad\qquad u_T(1) = I_X$$

we have $V_2V_1 = I_X$, hence we can apply Proposition 4.2 to the operator $u_T\colon A(D) \to B(X)$ and to $v = \pi_{|A(D)}$. Here again $A(D)$ is viewed as a subalgebra of the multiplication operators on $L_2(\mathbf{T})$, so that $A(D) \subset B(L_2(\mathbf{T}))$. It follows that the space E_1/E_2 appearing in Proposition 4.2 is a Hilbert space $(E_2 \subset E_1 \subset \hat{H})$ so that we obtain (ii). Conversely assume (ii). Let $\tilde{T} = STS^{-1}$. Since $\tilde{T} \in B(H)$ and $\|\tilde{T}\| \leq 1$ we have by Corollary 4.5 $\|u_{\tilde{T}}\|_{cb} \leq 1$ hence $\|u_T\|_{cb} \leq \|S\|\|S^{-1}\|\|u_{\tilde{T}}\|_{cb} \leq C$. This shows that (ii) $\Rightarrow$ (i) and concludes the proof. $\qquad\Box$

Note that (4.6) explains why the space X is necessarily isomorphic to a Hilbert space if u_T is completely bounded. Indeed, by Theorem 3.4, (4.6) implies that the identity on X factors through a Hilbert space with $\gamma_2(I_X) \leq \|u_T\|_{cb}$. The next result describes what happens if we remove this restrictive consequence of (4.6).

Corollary 4.14. *Let* X *be a Banach space. Let* $A_0 \subset A = A(D)$ *be the closure of the set of all polynomials* P *such that* $P(0) = 0$. *(Equivalently,* $A_0 = zA$ *or* $A_0 = \{f \in A \mid f(0) = 0\}$.) *Then* $u_{T|A_0}\colon A_0 \to B(X)$ *is completely bounded iff there is a factorization of* T *through a Hilbert space* H, *say* $T = ab$, $b\colon X \to H$, $a\colon H \to X$ *such that* $ba\colon H \to H$ *is a contraction, i.e.* $\|ba\| \leq 1$.

Proof: Assume first that $T = ab$ as above with $\|ba\| \leq 1$. Let $\tilde{T} = ba$. Observe the following crucial identity. For any polynomials Q and P with $P(z) = zQ(z)$, we have

$$P(\tilde{T}) = bQ(T)a \quad \text{and} \quad P(T) = aQ(\tilde{T})b.$$

It follows that $u_T(P) = au_{\tilde{T}}(Q)b$. This shows that u_T is c.b. on A_0 and

$$\|u_{T|A_0}\|_{cb} \leq \|a\|\|u_{\tilde{T}}\|_{cb}\|b\| \leq \|a\|\|b\|.$$

Conversely, assume that $\|u_{T|A_0}\|_{cb} \leq C$. Then, in particular since $u_T(z) = T$, by Theorem 3.4 we have $\gamma_2(T) \leq C$ hence we can factorize T through a Hilbert space H, i.e. we have $a_1\colon H \to X$ and $b_1\colon X \to H$ with $\|a_1\|\|b_1\| \leq C$ such that $T = a_1b_1$. Now let $T_1 = b_1a_1\colon H \to H$ and let P be an arbitrary polynomial. We can write $P = \alpha_0 + \alpha_1 z + zQ(z)$ where Q is a polynomial such that $Q(0) = 0$ and $\|Q\|_\infty \leq 3\|P\|_\infty$. Now

$$P(T_1) = \alpha_0 I_H + \alpha_1 T_1 + b_1 Q(T) a_1.$$

More generally for any matrix (P_{ij}) of polynomials we have

$$P_{ij}(T_1) = P_{ij}(0) I_H + P'_{ij}(0) T_1 + b_1 Q_{ij}(T) a_1$$

from which we immediately deduce

$$\|u_{T_1}\|_{cb} \le 1 + \|T_1\| + 3\|b_1\|\,\|a_1\|\,\|u_{T|A_0}\|_{cb}$$

and finally $\|u_{T_1}\|_{cb} \le 1 + C + 3C^2$.
But now Corollary 4.7 applies to T_1, and this yields an isomorphism $S\colon H \to H$ such that $\|S\|\,\|S^{-1}\| \le 1 + C + 3C^2$ and $\|S^{-1}T_1 S\| \le 1$. Taking $a = a_1 S$ and $b = S^{-1}b_1$ we obtain the desired conclusion with $\|a\|\,\|b\| \le C(1 + C + 3C^2)$. $\square$

Remark. In Pietsch's work on the spectral properties of Banach space operators ([Pie]), two operators $T_1\colon X \to X$ and $T_2\colon Y \to Y$ (on Banach spaces X and Y) are called *related* if there is a factorization of T_1 through Y, say $T_1 = ab$ with $b\colon X \to Y$, $a\colon Y \to X$ such that $T_2 = ba$. Pietsch repeatedly uses the fact that two related operators have the same non zero eigenvalues with the same multiplicity. With this terminology Corollary 4.14 gives a characterization of the operators on a Banach space which are related to a Hilbert space contraction. This opens the way to an extension of the Sz.-Nagy-Foias theory [SNF] to the Banach space case.

Problem. What property of T is equivalent to the complete boundedness of $u_{T|z^n A}\colon z^n A \to B(X)$ where $z^n A = \{f \in A \mid f(0) = f'(0) = \dots = f^{(n-1)}(0) = 0\}$? (This question has very recently been answered by V. Mascioni, [M]) More generally, we can ask the same question for a general ideal in A instead of $z^n A$ and also for ideals in H_∞.

We now give some complementary information on power bounded or on polynomially bounded operators.

First we mention without proof a remarkable estimate due to Bourgain [Bo].

Theorem 4.15. *Let $T\colon \ell_2^n \to \ell_2^n$ be an operator ($=$ an $n \times n$ matrix) such that $\|P(T)\| \le C\|P\|_\infty$ for all polynomials P. Then the associated homomorphism $u_T\colon P \to P(T)$ satisfies*

$$(4.7) \qquad \|u_T\|_{cb} \le K \, Log(n+1) C^4$$

where K is a numerical constant (independent of n in particular). Hence there is an isomorphism ($=$ a similarity) $S\colon \ell_2^n \to \ell_2^n$ such that

$$(4.7)' \qquad \|S^{-1}TS\| \le 1 \quad and \quad \|S\|\,\|S^{-1}\| \le K \, Log(n+1) C^4.$$

Of course (4.7) implies (4.7)' by Paulsen's criterion (Corollary 4.7) but Bourgain's proof of (4.7) is rather complicated.
To state the next result, we simplify the notation. We denote ($1 \le p \le \infty$)

$$A = \{f \in C(\mathbf{T}) \mid \hat{f}(n) = 0 \;\forall n < 0\}, \quad L_p = L_p(\mathbf{T}, m), \quad L_\infty = L_\infty(\mathbf{T}, m) \quad and$$

$$H_p = \{f \in L_p(\mathbf{T}, m) \mid \hat{f}(n) = 0 \;\forall n < 0\}.$$

More generally, for any Banach space X and for $1 \leq p < \infty$, we will denote by $L_p(X)$ the Banach space of all X-valued Bochner-measurable functions $f \colon \mathbf{T} \to X$, which are p-integrable with respect to the normalized Haar measure m on $\mathbf{T}$, and we equip it with its natural norm. Then we again define

$$(4.8) \qquad H_p(X) = \{f \in L_p(X) \mid \hat{f}(n) = 0 \; \forall n < 0\},$$

and we equip it with the induced norm.

We will consider the convolutions $F * \varphi$ with $F \in H_1$, $\varphi \in L_\infty$. Note that $F * \varphi(e^{it}) = \sum_{k \geq 0} \hat{F}(k)\hat{\varphi}(k)e^{ikt}$. By convention, we will write $F(T) = \sum_{k \geq 0} \hat{F}(k)T^k$ and in particular

$$F * \varphi(T) = \sum_{k \geq 0} \hat{F}(k)\hat{\varphi}(k)T^k$$

whenever it makes sense (for instance if $F * \varphi$ is a polynomial).

The following rather simple estimate and its corollary are useful. These are due to Peller [Pe1] (for $n = 1$) and Bourgain [Bo]. (Note however that we only use the Cauchy-Schwarz inequality, while Bourgain [Bo] uses Grothendieck's inequality to prove this.)

Proposition 4.16. *Let* $T \colon H \to H$ *be an arbitrary operator. Consider F in H_1, and an $n \times n$ matrix (φ_{ij}) with entries in L_∞. Then if (say) $F * \varphi_{ij}$ is a polynomial for all i, j we have*

$$(4.9) \qquad \|(F * \varphi_{ij}(T))\|_{M_n(B(H))} \leq \|F\|_{H_1} \sup_{k \geq 0} \|T^k\|^2 \|(\varphi_{ij})\|_{L_\infty(M_n)}$$

where we have defined

$$\|(\varphi_{ij})\|_{L_\infty(M_n)} = \operatorname*{ess\,sup}_{t \in \mathbf{T}} \|(\varphi_{ij}(e^{it}))\|_{M_n}.$$

Proof: The proof is easy using the fact that every H_1 function F can be written as a product $F = gh$ with $g, h \in H_2$ such that $\|g\|_2 \|h\|_2 = \|F\|_{H_1}$. Note that we clearly have if $F = gh$ (we leave some minor technical details to the reader)

$$F * \varphi_{ij}(T) = \int F(e^{-is}T)\varphi_{ij}(e^{is})dm(s) = \int g(e^{-is}T)h(e^{-is}T)\varphi_{ij}(e^{is})dm(s).$$

Now consider $x_j \in H$, $y_i \in H$ such that $\sum \|x_j\|^2 \leq 1$ and $\sum \|y_i\|^2 \leq 1$. We have

$$\sum_{ij} \langle F * \varphi_{ij}(T)x_j, y_i \rangle = \int \sum_{ij} \langle h(e^{-is}T)x_j, g(e^{-is}T)^* y_i \rangle \varphi_{ij}(e^{is})dm(s).$$

Hence if $\|(\varphi_{ij})\|_{L_\infty(M_n)} \leq 1$, this is less than (or equal to)

$$\int \left(\sum_j \|h(e^{-is}T)x_j\|^2 \right)^{1/2} \left(\sum_i \|g(e^{-is}T)^* y_i\|^2 \right)^{1/2} dm(s)$$

which by Cauchy-Schwarz is itself less than (or equal to)

$$\left(\sum_j \int \|h(e^{-is}T)x_j\|^2 dm(s)\right)^{1/2} \left(\sum_i \int \|g(e^{-is}T)^* y_i\|^2 dm(s)\right)^{1/2}$$

$$= \left(\sum_j \sum_{k\geq 0} |\hat{h}(k)|^2 \|T^k x_j\|^2\right)^{1/2} \left(\sum_i \sum_{k\geq 0} |\hat{g}(k)|^2 \|T^{*k} y_i\|^2\right)^{1/2}$$

$$\leq (\sup \|T^k\|)^2 \|g\|_2 \|h\|_2 \left(\sum_j \|x_j\|^2 \sum_i \|y_i\|^2\right)^{1/2} \leq (\sup \|T^k\|)^2 \|F\|_{H_1}.$$

By homogeneity this completes the proof of (4.9). □

Corollary 4.17. *Let $T: H \to H$ be a power bounded operator and let (P_{ij}) be a matrix of polynomials. Let $d = \max_{ij} \deg(P_{ij})$. Then we have*

$$\|(P_{ij}(T))\|_{M_n(B(H))} \leq K \, Log(1+d) \sup_{|z|=1} \|(P_{ij}(z))\|_{M_n} (\sup_{k\geq 1} \|T^k\|)^2$$

where K is a numerical constant.

Proof: Let $F = \sum_{0 \leq k \leq d} e^{ikt}$. Then it is well known that $\|F\|_{H_1} \leq K \, Log(1+d)$ for some constant K independent of d, and since $P_{ij} = P_{ij} * F$, this follows from (4.9). □

Recently, Daher [Da] observed that Bourgain's proof of Theorem 4.15 actually proves quite a bit more. To explain this we need more notation. Let T be a polynomially bounded operator in $B(H)$ and let $u_T: A(D) \to B(H)$ be the associated homomorphism. Let $\{T\}'$ denote as usual the commutant of T in $B(H)$. Note that this is a unital (in general non self adjoint) subalgebra of $B(H)$. We may consider the map

$$\hat{u}_T: A(D) \otimes \{T\}' \longrightarrow B(H)$$

defined (on the algebraic tensor product) by

$$\hat{u}_T(f \otimes b) = f(T)b \; (= bf(T)).$$

Equivalently $\hat{u}_T$ maps any polynomial $P = \sum_0^n a_k z^k$ with coefficients a_k in $\{T\}'$ to the operator $\hat{u}_T(P) = \sum a_k T^k = \sum T^k a_k$.

Definition. *We will say that u is strongly polynomially bounded (resp. strongly completely polynomially bounded) if the unital homomorphism $\hat{u}_T$ extends to a bounded (resp. completely bounded) homomorphism on the closure of $A(D) \otimes \{T\}'$ in the space $C(\overline{D}; B(H))$. We denote this closure by $A(D, \{T\}')$.*

Moreover, we will denote by $C^s(T)$ (resp. $C_{cb}^s(T)$) the norm (resp. the c.b. norm) of $\hat{u}_T\colon A(D,\{T\}') \to B(H)$.

By Theorem 4.8, $\|\hat{u}_T\|_{cb} \le 1$ iff T admits a unitary dilation U on $\widehat{H} \supset H$ for which there is a $*$-representation $\pi\colon B(H) \to B(\widehat{H})$ with range commuting with U such that

$$\forall n \ge 1 \quad \forall a \in \{T\}'. \qquad\qquad aT^n = P_H \pi(a) U_{|H}^n.$$

In other words, we can somewhat simultaneously and coherently dilate T and all operators commuting with T.

It should be emphasized that the elements of $\{T\}'$ do not in general commute with T^*, and in general a contraction T does *not* satisfy $C^s(T) \le 1$. However, if T is unitary (or if T is a normal contraction using Fuglede's well known theorem [Fug or Ha2 p.68]) it is true and we have even $C_{cb}^s(T) \le 1$. (We leave this as an exercise for the reader.)

Precisely, what Daher [Da] observed is the following statement (the proof is identical to the one in Bourgain's paper [Bo]).

Theorem 4.18. *Let $T\colon \ell_2^n \to \ell_2^n$ be polynomially bounded with constant C. Then*

$$C_{cb}^s(T) = \|\hat{u}_T\|_{cb} \le K \ Log(n+1)C^4$$

for some numerical constant K.

Remark. Note that (unlike Bourgain's original formulation) this is already non-trivial for a contraction! This suggests the following.

Problem 4.19. Estimate the number

$$\sup\{C_{cb}^s(T) \mid T \in M_n, \ \|T\| \le 1\}$$

when $n \to \infty$.
Note that by Lemma 4.20 below and Proposition 1.6, this number is > 1 for all n large enough. By the example preceding Proposition 1.6, $n \ge 8$ suffices, and by the remark after Corollary 5.9 below, $n \ge 6$ suffices. We suspect that $Log(n+1)$ is asymptotically sharp in Theorem 4.18.

The main advantage of the constants $C^s(T)$ and $C_{cb}^s(T)$ is that they permit an iteration, as follows.

Lemma 4.20. *Let $\tau = (T_1, \ldots, T_k)$ be a k-tuple of mutually commuting operators in $B(H)$. Assume $C^s(T_i) < \infty$ (resp. $C_{cb}^s(T_i) < \infty$) for each $i = 1, \ldots, k$. Then the homomorphism*

$$u_\tau\colon A(D^k) \to B(H)$$

which takes the i-th coordinate z_i on D^k to T_i is well defined and bounded. Moreover, we have

$$\|u_\tau\| \le C^s(T_1) \ldots C^s(T_k)$$

$$(resp. \ \|u_\tau\|_{cb} \le C_{cb}^s(T_1) \ldots C_{cb}^s(T_k)).$$

Proof: We use induction on k. The statement is trivial for $k = 1$. Assume that we have proved this for any $(k-1)$-tuple and let us check it for a k-tuple. Consider a polynomial P in $A(D^k)$. We can write

$$P = \sum_{n \geq 0} z_k^n P_n(z_1, \ldots, z_{k-1})$$

where P_n is a polynomial in $z_1, \ldots, z_{k-1}$ and the sum is actually finite. By definition of $C_s(T_k)$ we have

$$\left\| \sum_{n \geq 0} T_k^n P_n(T_1, \ldots, T_{k-1}) \right\| \leq C_s(T_k) \sup_{|z_k|=1} \left\| \sum z_k^n P_n(T_1, \ldots, T_{k-1}) \right\|$$

but by the induction hypothesis, for each fixed z_k, we have

$$\left\| \sum z_k^n P_n(T_1, \ldots, T_{k-1}) \right\| \leq C^s(T_1) \ldots C^s(T_{k-1})$$

$$\sup \left\{ \left\| \sum z_k^n P_n(z_1, \ldots, z_{k-1}) \right\| \ \middle| \ (z_1, \ldots, z_{k-1}) \in D^{k-1} \right\},$$

hence we obtain

$$\|P(T_1, \ldots, T_k)\| = \left\| \sum T_k^n P_n(T_1, \ldots, T_{k-1}) \right\| \leq C^s(T_1) \ldots C^s(T_k) \|P\|_\infty.$$

The proof for C_{cb}^s is the same (left to the reader).   □

As Daher noticed, we have the following interesting application.

Corollary 4.21. *Let $\tau = (T_1, \ldots, T_k)$ be k mutually commuting contractions in M_n (resp. assume each T_i polynomially bounded with constant C_i, $i = 1, \ldots, k$). Then the homomorphism*

$$u_\tau \colon A(D^k) \longrightarrow M_n$$

which takes z_i to T_i satisfies

$$\|u_\tau\|_{cb} \leq (K \, Log(n+1))^k$$

(resp. $\|u_\tau\|_{cb} \leq (K \, Log(n+1))^k (C_1 C_2 \ldots C_k)^4$).

Proof: This is an immediate consequence of Theorem 4.18 and Lemma 4.20.   □

Remark. Let $\tau = (T_1, \ldots, T_k)$ be as in Corollary 4.21. Then, by Theorem 4.3, there is a similarity $S \colon \ell_2^n \to \ell_2^n$ with $\|S^{-1}\| \|S\| \leq \|u_\tau\|_{cb}$ such that the k-tuple $(S^{-1}T_1 S, \ldots, S^{-1}T_k S)$ can be dilated to a k-tuple of mutually commuting unitaries.

It would be interesting to understand which operator algebras or which uniform algebras A have the property that every bounded homomorphism $u \colon A \to B(H)$ is c.b. By "operator algebra", we mean here a closed subalgebra of $B(\mathcal{H})$ for some Hilbert space $\mathcal{H}$. On the other hand, by a uniform algebra we mean a closed unital subalgebra of $C(T)$ (for some compact set T) which separates the points of T. Since $C(T)$ can be realized as a commutative unital

C^*-subalgebra of $B(H)$, it is an operator algebra. Hence, any uniform algebra can be viewed as an operator algebra.

Recall that by Remark 3.20, the various possible realizations of $C(T)$ as a C^*-subalgebra of $B(H)$, hence the various realizations of A as an operator algebra, yield the same notion of complete boundedness for maps defined on A (or into A). For instance, let $\Omega \subset \mathbb{C}^n$ be a bounded open connected subset. We denote by $A(\Omega)$ the closure of the polynomials in the uniform norm, so that $A(\Omega)$ is a uniform algebra embedded isometrically into $C(\partial\Omega)$. A fortiori, we will view $A(\Omega)$ as an operator algebra.

Now let $A \subset B(\mathcal{H})$ be an arbitrary operator algebra. We wish to discuss the following two properties of A:

(B) Every bounded unital homomorphism $u\colon A \to B(H)$ (H arbitrary Hilbert) is $c.b.$

(C) Every contractive unital homomorphism $u\colon A \to B(H)$ is completely contractive.

Of course, we already know that $A(D)$ and $A(D^2)$ have property (C). It is unknown (Problem 0.3) whether the disc algebra $A(D)$ has property (B). Actually, quite shockingly there seems to be no nontrivial example of a uniform algebra of the form $A(\Omega)$ ($\Omega \subset \mathbb{C}^n$) which either satisfies or fails property (B). (Here the trivial case is when $A(\Omega) = C(\partial\Omega)$. Then, $C(\partial\Omega)$ being a nuclear C^*-algebra, property (B) follows e.g. from Theorem 7.16 below.)

However, there are known examples failing property (C). For instance, Parrott's construction in [Par1] implies that, if $\Omega = D^3$, then $A(\Omega)$ fails property (C). There are also fairly simple examples of uniform algebras failing property (B), but over infinite dimensional domains Ω, as we will now see. To produce these, we follow an idea originating in Misra and Satry's papers [MS1-2] and developed by Paulsen in [Pa5]. We first describe a brutal way to produce nonunital counterexamples to property (B). Consider a closed subspace $E \subset B(H)$. Then consider the subalgebra $A_E \subset B(H \oplus H)$ formed of all elements of the form $\begin{pmatrix} 0 & x \\ 0 & 0 \end{pmatrix}$ with $x \in E$. Equivalently, A_E can be viewed as the algebra obtained by equipping E with the identically zero product operation, namely we have $x \cdot y = 0$ for all x, y in A_E. Moreover, E and A_E are clearly completely isometric. Observe that any linear mapping $v\colon E \to B(\mathcal{H})$ defines a homomorphism $\hat{v}\colon A_E \to B(\mathcal{H} \oplus \mathcal{H})$ by setting $\hat{v}\begin{pmatrix} 0 & x \\ 0 & 0 \end{pmatrix} = \begin{pmatrix} 0 & v(x) \\ 0 & 0 \end{pmatrix}$. Moreover, we have $\|v\| = \|\hat{v}\|$ and $\|v\|_{cb} = \|\hat{v}\|_{cb}$. Therefore, in this case, if every bounded homomorphism $u\colon A_E \to B(\mathcal{H})$ is $c.b.$ then any bounded *linear* map $v\colon E \to B(H)$ is $c.b.$, but this is not true in general (see (3.17), Theorem 3.21 or Remark 3.23). Hence, we easily obtain operator algebras of the form A_E which fail property (B).

To obtain analogous unital examples, we simply replace A_E by the closed span of A_E and the identity $\begin{pmatrix} I & 0 \\ 0 & I \end{pmatrix}$.

We now consider a finite dimensional normed space E. We can assume $E = \mathbb{C}^n$ as a vector space. Let $\Omega_E \subset \mathbb{C}^n$ be the open unit ball of the dual space E^*. We will show that, if E is suitably chosen then $A(\Omega_E)$ fails property (C). This follows from the next result.

Theorem 4.22. *Let $\beta > 0$ be a constant.*

(i) *Assume that for any contractive unital homomorphism $u \colon A(\Omega_E) \to B(H)$ we have $\|u\|_{cb} \leq \beta$. Then the constant $\alpha(E)$ defined in (3.21) satisfies*

$$\alpha(E) \leq \beta.$$

(ii) *In particular, if $A(\Omega_E)$ has property (C) then we have $\alpha(E) = 1$.*

(iii) *If $E = \ell_2^n$ and $n > 1$, then $A(\Omega_E)$ fails property (C).*

Proof. Let $w \colon E \to B(H)$ be a linear map with $\|w\| \leq 1$. We will show that $\|w\|_{cb} \leq \beta$ when we consider E as embedded into $C(\overline{\Omega}_E)$. Let $e_1, \ldots, e_n$ be the canonical basis of $\mathbb{C}^n$ and let $T_i = w(e_i)$. We introduce the mapping $u \colon A(\Omega_E) \to B(H \oplus H)$ defined by

$$\forall f \in A(\Omega_E) \qquad u(f) = \begin{pmatrix} f(0)I_H & \sum\limits_{i=1}^{n} \frac{\partial f}{\partial z_i}(0)T_i \\ 0 & f(0)I_H \end{pmatrix}.$$

Evidently u is a unital homomorphism.
We will use the following elementary inequality for any f in $A(\Omega_E)$

$$(4.10) \qquad \left\| \sum_{1}^{n} \frac{\partial f}{\partial z_i}(0)e_i \right\|_E \leq \|f\|.$$

This is proved as follows: let f be a polynomial, let

$$f = \sum_{m \geq 0} f_m$$

be the decomposition of f into a sum of homogenous polynomial with f_m of degree m. Note $f_0(z) \equiv f(0)$ and $f_1(z) = \sum\limits_{1}^{n} \frac{\partial f}{\partial z_i}(0)z_i$. Since $f(e^{it}z) = \sum\limits_{m \geq 0} e^{imt} f_m(z)$ we have

$$f_1(z) = \int e^{-it} f(e^{it}z) \frac{dt}{2\pi}$$

hence

$$\left\| \sum \frac{\partial f}{\partial z_i}(0)e_i \right\|_E = \sup_{z \in \Omega_E} \left\| \sum \frac{\partial f}{\partial z_i}(0)z_i \right\| = \|f_1\| \leq \|f\|.$$

This establishes (4.10). Now we claim that $\|u(f)\| \leq \|f\|$ for all f in $A(\Omega_E)$. Indeed, assume first that $f(0) = 0$. Then

$$\|u(f)\| = \left\|\sum \frac{\partial f}{\partial z_i}(0)T_i\right\| = \left\|w\left(\sum \frac{\partial f}{\partial z_i}(0)e_i\right)\right\| \le \left\|\sum \frac{\partial f}{\partial z_i}(0)e_i\right\|_E$$

whence the desired result by (4.10). For the general case assume $\|f\| < 1$. Consider a Möbius transformation φ on the unit disc such that $\varphi(f(0)) = 0$. By the preceding argument we have $\|u(\varphi \circ f)\| \le 1$. But now, if $T = u(\varphi \circ f)$, T is a contraction, hence by von Neumann's inequality, we have $\|\varphi^{-1}(T)\| \le 1$ and since u is a homomorphism we easily verify $\varphi^{-1}(u(\varphi \circ f)) = u[\varphi^{-1}(\varphi \circ f)] = u(f)$. Thus we obtain $\|u\| \le 1$, as claimed above. By our assumption this implies

$$\|u\|_{cb} \le \beta.$$

Now let $J: E \to A(\Omega_E)$ be the natural isometric embedding which takes e_i to the i-th coordinate z_i restricted to Ω_E. Invoking Remark 3.20 and the embedding $A(\Omega_E) \subset C(\partial\Omega_E)$ the cb-norm of w makes good sense and is the same whether we use J or the natural embedding $E \subset C(\overline{\Omega}_E)$. Moreover, we have for any e in E

$$w(e) = P_1 u J(e) P_2^*$$

where $P_k: H \oplus H \to H$ is the canonical projection onto the k-th coordinate, $k = 1, 2$.

Therefore we conclude that

$$\|w\|_{cb} \le \|uJ\|_{cb} \le \|u\|_{cb} \le \beta.$$

This shows that $\alpha(E) \le \beta$. The assertion (ii) is clear. To verify (iii), simply recall that by Remark 3.22 we have $\alpha(\ell_2^n) > 1$ for any $n > 1$. ☐

Remark 4.23. Consider the 2-dimensional version of ℓ_1, namely the space $E = \ell_1^2$. Then by Ando's dilation theorem, the algebra $A(\Omega_E) = A(D^2)$ possesses property (C). This shows that $\alpha(\ell_1^2) = 1$. One can also check $\alpha(\ell_\infty^2) = 1$. It is unknown (see [Pa5]) whether these are the only spaces E (of dimension > 1) with $\alpha(E) = 1$. Note however that by Remark 3.24, $\alpha(E) = 1$ implies $\dim E \le 4$.

We refer the reader to [AFJS] for some closely related results.

Notes and Remarks on Chapter 4

Proposition 4.1 is a variant on a theme initiated by Sarason [Sa] (see Theorem 1.7) who recognized the crucial rôle played by the notion of semi-invariance and by compressions in dilation theory. These ideas were extended to the Banach space case in [P4]. (Propositions 4.1 and 4.2 were essentially observed in [P4].) Theorem 4.3 is due to Paulsen [Pa2-3-4], who was mainly motivated by the case when the algebra A is the disc algebra (i.e. by the non self-adjoint case). The case when A is a C^*-algebra (i.e. Corollary 4.4) had been previously settled on one hand by a combination of the works of Hadwin [Had] and Wittstock [Wi1], and on the other independently by Haagerup [H1]. Actually in the C^*-case, the precise equality of $\|u\|_{cb}$ in Theorem 4.3 with the "similarity constant" $\min\{\|S\|\,\|S^{-1}\|\}$ was only obtained in [H1]; the Hadwin-Wittstock approach via decomposition into completely positive maps yields the equivalence of these two parameters with an extra numerical factor.

Lemma 4.5 seems to be a well known fact. For its proof, we have followed [Pa1]. Corollary 4.6 was first obtained by Christensen [C2]. It can also be deduced from [H1] (this is what we do in the text above).

Concerning Corollary 4.6, there is an extensive literature on numerous ramifications of the derivation problem, cf. e.g. [J, Ri, K4, S]. For instance, it is known that every derivation $\delta\colon M \to M$ defined on a von Neumann algebra M (and with values into M) is inner. This is a classical result due to Kadison [K4] and Sakai [S]. Moreover, it is known (cf. e.g. [SS, p. 62]) that any derivation $\delta\colon A \to A$ defined on a C^*-algebra $A \subset B(H)$ extends to a derivation defined on the von Neumann algebra generated by A. Therefore, in numerous situations, it suffices to consider the von Neumann algebra case. See [Di2, Chapter III, §9] for a rapid exposition. For a more recent treatment (including multilinear extensions) see [SS].

It is natural to ask for an analogue of the identity appearing in Corollary 4.4, for derivations. Here is the answer, at least in the von Neumann algebra case: Let $M \subset B(H)$ be a von Neumann subalgebra and let $\delta\colon M \to B(H)$ be a c.b. derivation (hence inner by Corollary 4.6). Then we have (recall that δ_T is defined by $\delta_T(x) = Tx - xT$)

$$\|\delta\|_{cb} = 2\inf\{\|T\| \mid \delta = \delta_T\}$$

or equivalently for any inner derivation δ_T we have

$$(4.11) \qquad \|\delta_T\|_{cb} = 2\inf\{\|T - \widehat{T}\| \mid \widehat{T} \in M'\}$$

where M' denotes the commutant of M. We will denote by $d(T, M')$ the right side of (4.11). Here is a quick proof of (4.11) taken from [C5] (but the argument is due to Arveson, this is closely related to Arveson's celebrated distance formula, see [Ar2, p.12]). We will prove that

$$(4.12) \qquad d(T, M') \le (1/2)\|\delta_T\|_{cb}.$$

First note that $\|\delta_T\|_{cb}$ is equal to the norm of the operator $I_{B(\ell_2)} \otimes \delta_T$ acting on the (von Neumann algebraic) tensor product $B(\ell_2)\overline{\otimes}M \subset B(\ell_2(H))$. Let $B(H)_*$ denote the predual of $B(H)$. By Hahn-Banach, we have

$$(4.13) \qquad d(T, M') = \sup\{|f(T)| \mid f \in B(H)_*, f \perp M', \|f\| = 1\}.$$

Consider $f \in B(H)_*$ with $f \perp M'$ and $\|f\| = 1$. By the classical identification of $B(H)_*$ with the space of trace class operators on H, there are unit vectors ξ, η in $\ell_2(H)$ such that

$$(4.14) \qquad \forall x \in B(H) \quad f(x) = \langle (I \otimes x)\xi, \eta \rangle.$$

Hence for any x in M', $(I \otimes x)\xi$ is orthogonal to η. Let p be the orthogonal projection on $\ell_2(H)$ onto the closure of $\{(I \otimes x)\xi \mid x \in M'\} \subset \ell_2(H)$. Clearly $p\xi = \xi$ and $p\eta = 0$. Moreover, since p commutes with $(I \otimes M')$, we have $p \in B(\ell_2)\overline{\otimes}M$. By (4.14), we can write

$$|f(T)| = |\langle (I \otimes T)p\xi, (1 - p)\eta \rangle|$$
$$\le \|(1 - p)(I \otimes T)p\| = \|[(I \otimes T)p - p(I \otimes T)]p\|$$
$$\le \|(I \otimes T)p - p(I \otimes T)\|$$
$$= \frac{1}{2}\|(I \otimes T)(2p - 1) - (2p - 1)(I \otimes T)\| = \frac{1}{2}\|(I \otimes \delta_T)(2p - 1)\|$$

hence (note that $2p - 1 = p - (1 - p)$ is unitary)

$$\le \frac{1}{2}\|I \otimes \delta_T\|.$$

Taking the supremum over all possible f and using (4.13), we obtain (4.12). The converse is obvious.   □

As already mentioned, (4.11) is closely related to Arveson's distance formula for nest algebras. In [Ar2], a subalgebra $M \subset B(H)$ is called hyper-reflexive if there exists a constant K such that for any T in $B(H)$ we have

$$(4.15) \qquad d(T, M) \le K \sup\{\|(I - P)TP\|\}$$

where the supremum runs over the family $\mathrm{lat}(M)$ of all self-adjoint projections P onto M-invariant subspaces, or equivalently such that $(I - P)xP = 0$ for all x in M. For the record, M is called reflexive if T belongs to M whenever the right side of (4.15) vanishes.

We claim that if M is a von Neumann subalgebra of $B(H)$ and every bounded derivation $\delta: M' \to B(H)$ is inner (or equivalently c.b. by Corollary 4.6), then M is hyper-reflexive.

Indeed, note that when M is self-adjoint, then $P \in \text{lat}(M)$ iff $1 - P \in \text{lat}(M)$. Moreover, observe $\|TP - PT\| \leq \|TP - PTP\| + \|PTP - PT\|$. Therefore, if M is a von Neumann algebra, we have

$$(4.16) \qquad \|\delta_T\|_{M' \to B(H)} \leq 4 \sup_{P \in M'} \{\|TP - PT\|\} \leq 8 \sup\{\|(1 - P)TP\|\}$$

where the last supremum is as above. Now if every bounded derivation $\delta \colon M' \to B(H)$ is $c.b.$, then (by the closed graph theorem) there is obviously a constant K' such that any such derivation satisfies $\|\delta\|_{cb} \leq K'\|\delta\|$, which implies by (4.16) and (4.12) that $M = M''$ is hyper-reflexive.

Thus, the problem whether every von Neumann algebra is hyper-reflexive, which is a well known open problem in Arveson's theory (see the lectures 7 and 8 in [Ar2]), is closely related to the derivation problem (Problem 0.2').

For more information and recent work on the cb-norm of a derivation, see [Mat]. For a derivation $\delta \colon M \to M$ (acting *into* M), actually it is known that $\|\delta\| = \|\delta\|_{cb}$, hence (4.11) holds with the norm instead of the cb-norm. This was first proved in [St] (for $M = B(H)$) and in [Zs] (for M arbitrary)

Corollary 4.7 is the main result of [Pa3]. Its Banach space version (Corollary 4.13) comes from [P4]. Along the same lines as [P4], Corollary 4.14 is new. See Mascioni's paper [M] for further developments. Statements 4.8 and 4.9 are well known consequences of Arveson's ideas in [Ar1]. Corollaries 4.10 and 4.11 are due to Rota [Ro], while corollary 4.12 is due to B. Sz.-Nagy [SN].

Theorem 4.15 is due to Bourgain [Bo]. It is of course open whether the factor $Log(n + 1)$ can be removed from this estimate. While it is widely believed that this would essentially solve the similarity Problem 0.3, there does not seem to be a rigorous argument! The converse, however, is quite clear (by a standard direct sum argument): if Problem 0.3 has an affirmative answer, then the $Log(n + 1)$ factor can be deleted from (4.7), so that (4.7) would then become $\|u_T\|_{cb} \leq f(C)$ for some function of C. There is some known information on the constants related to a possible solution of this problem, for instance (4.7) cannot be replaced by $\|u_T\|_{cb} \leq C$, see [Ho1-2-3]. We refer the reader to a forthcoming paper of ours for more on this.

Using a new dilation theorem, the similarity problem for polynomially bounded (in short p.b.) operators is reduced in [Pet] to the special case of those p.b. operators T which are also "weakly centered", *i.e.* such that T^*T commutes with TT^*. In addition one may assume that the spectrum of T is the unit circle. Related results appear in [PPP].

The paper [Pe1] contains numerous facts on power bounded operators, in particular Proposition 4.16 for $n = 1$ and Corollary 4.17. We refer the reader to §6 in [Pe1] for a thorough discussion of Proposition 4.16 and its possible optimality in some sense. See the notes and remarks on chapter 6 in this volume for further comments on Peller's paper.

Theorem 4.18, Lemma 4.20 and Corollary 4.21 are all based on [Da].

Theorem 4.22 is due to Paulsen [Pa5]. We refer the reader to [DP, CCFW] for the related theory of Hilbert modules over function algebras such as for instance $A(\Omega)$.

5. Schur multipliers and Grothendieck's inequality

Summary: In this chapter, we study Schur multipliers on the space $B(H,K)$ of all bounded operators between two Hilbert spaces. We give a basic characterization of the unit ball of the space of Schur multipliers, in connection with the class of operators factoring through a Hilbert space (considered above in chapter 3). Then we prove Grothendieck's fundamental theorem (= Grothendieck's inequality) in terms of Schur multipliers. We give Varopoulos's proof that, since the Grothendieck constant is > 1, Ando's inequality does not extend with constant 1 to n-tuples of mutually commuting contractions. Finally, we discuss the extensions to Schur multipliers acting boundedly on the space $B(H,K)$ when H,K are replaced by ℓ_p-spaces, $1 \leq p < \infty$.

Let S, T be arbitrary sets. We will denote by $\ell_p(T)$ the space of all complex valued functions α: $t \to \alpha_t$ such that $\sum |\alpha_t|^p < \infty$ with the usual norm $\|\alpha\|_p = \left(\sum |\alpha_t|^p \right)^{1/p}$. If $p = \infty$, we let $\|\alpha\|_\infty = \sup |\alpha_t|$. We denote by $c_0(T)$ the subspace of $\ell_\infty(T)$ formed by the sequences tending to zero at infinity. We equip $c_0(T)$ with the norm $\| \ \|_\infty$ induced by $\ell_\infty(T)$.

A function φ: $S \times T \to \mathbb{C}$ is called a Schur multiplier on $B(\ell_2(T), \ell_2(S))$ if for every operator A: $\ell_2(T) \to \ell_2(S)$ with associated matrix $(a(s,t))$ the matrix $(\varphi(s,t)a(s,t))$ also represents a bounded operator from $\ell_2(T)$ into $\ell_2(S)$. We then denote $M_\varphi A = (\varphi(s,t)a(s,t))$. Clearly, if φ is a Schur multiplier on $B(\ell_2(T), \ell_2(S))$, the associated linear operator

$$M_\varphi \colon B(\ell_2(T), \ell_2(S)) \to B(\ell_2(T), \ell_2(S))$$

is necessarily bounded.

The following result essentially due to Grothendieck [G] is fundamental.

Theorem 5.1. *Let φ: $S \times T \to \mathbb{C}$ be a function and let $C \geq 0$ be a constant. The following are equivalent.*

(i) *φ is a bounded Schur multiplier on $B(\ell_2(T), \ell_2(S))$ with norm $\|M_\varphi\| \leq C$.*
(ii) *There is a Hilbert space H and families $(y_s)_{s \in S}$, $(x_t)_{t \in T}$ of elements of H such that*

$$\forall\, s \in S, \forall\, t \in T \qquad \varphi(s,t) = \langle x_t, y_s \rangle \quad \text{and} \quad \sup_{s \in S} \|y_s\| \sup_{t \in T} \|x_t\| \leq C.$$

(iii) *The operator u_φ: $\ell_1(T) \to \ell_\infty(S)$ which admits $(\varphi(s,t))$ as its matrix is in the space $\Gamma_2(\ell_1(T), \ell_\infty(S))$ and $\gamma_2(u_\varphi) \leq C$.*

(iv) M_φ is completely bounded on $B(\ell_2(T), \ell_2(S))$ with $\|M_\varphi\|_{cb} \leq C$.

The proof is a simple consequence of a basic property of the γ_2-norm, which we now describe. Here (e_t) denotes the canonical basis in $\ell_1(T)$.

Lemma 5.2. *Let $(y_i)_{i \leq n}$ be a finite sequence of finitely supported elements in the Banach space $\ell_1(T)$. The following are equivalent.*

(i) *There is a finite sequence (x_j) in $\ell_1(T)$ with $\sum \|x_j\|^2 \leq 1$ such that $(y_i) < (x_j)$ (in the sense of (3.1))*

(ii) *There is a finitely supported sequence of scalars $(\beta_t)_{t \in T}$ with $\sum |\beta_t|^2 \leq 1$ and an operator $A: \ell_2(T) \to \ell_2^n$ with $\|A\| \leq 1$ such that*

$$y_i = \sum_{t \in T} a(i,t)\beta_t e_t.$$

(iii) *There is a finite sequence (x_j) of the form $x_j = \beta_j e_{t_j}$ with t_j distinct and such that*

$$\sum |\beta_j|^2 \leq 1 \quad \text{and} \quad t_j \in T \quad \text{such that} \quad (y_i) < (x_j).$$

Proof: Note that (ii) and (iii) are clearly equivalent by Lemma 3.3 and (iii) $\Rightarrow$ (i) is obvious. It remains to prove (i) $\Rightarrow$ (ii). This is quite simple. Assume (i). We may assume $x_j \neq 0$. Let $x_j' = x_j(\|x_j\|^{-1})$ so that $\sum_t |x_j'(t)| = 1$. We have then for all ξ in $c_0(T)$ by convexity

$$\sum_j |\sum_t x_j(t)\xi(t)|^2 \leq \sum_j \|x_j\|^2 |\sum_t x_j'(t)\xi(t)|^2$$

$$\leq \sum_j \|x_j\|^2 \sum_t |x_j'(t)||\xi(t)|^2$$

$$\leq \sum_t \beta_t^2 |\xi(t)|^2$$

where we have set $\beta_t = (\sum_j \|x_j\|^2 |x_j'(t)|)^{1/2}$.

Since $(y_i) < (x_j)$, we have $\sum_i |\sum_t y_i(t)\xi(t)|^2 \leq \sum_j |\sum_t x_j(t)\xi(t)|^2$, hence for all $\xi \in c_0(T)$

$$\sum_i |\sum_t y_i(t)\xi(t)|^2 \leq \sum_t \beta_t^2 |\xi(t)|^2.$$

Equivalently this means $(y_i) < (\beta_t e_t)_{t \in T}$. Hence, there exists an operator $A : \ell_2(T) \to \ell_2^n$ with $\|A\| \leq 1$ and moreover such that

$$\forall \xi \in c_0(T) \quad \sum_t y_i(t)\xi(t) = \sum_t a(i,t)\beta_t \xi(t).$$

In other words, $y_i = \sum_t a(i,t)\beta_t e_t$. Note $\sum |\beta_t|^2 \leq 1$. Moreover, since we assume (y_i) finitely supported, we may as well replace T by the union of the supports of the elements (y_i) and assume that T itself is finite. Then we obtain (ii). $\quad\square$

Let E, F be Banach spaces. On $E \otimes F$ we introduce the tensor norm γ_2^* defined as follows. For all v in $E \otimes F$ we define

$$(5.1) \qquad \gamma_2^*(v) = \inf \left\{ \left(\sum \|x_j\|^2 \right)^{1/2} \left(\sum \|\xi_i\|^2 \right)^{1/2} \right\}$$

where the infimum runs over all finite sequences (y_i), (x_j) in E and (ξ_i) in F such that $(y_i) < (x_j)$ and $v = \sum y_i \otimes \xi_i$. It is not hard to check that γ_2^* is a norm on $E \otimes F$. We will denote by $E \widehat{\otimes}_{\gamma_2^*} F$ the completion of $E \otimes F$ with respect to that norm. The next result is but a reformulation of Theorem 3.4.

Theorem 5.3. *Let E and F be arbitrary Banach spaces. Then we have a natural isometric isomorphism*

$$(E \widehat{\otimes}_{\gamma_2^*} F)^* = \Gamma_2(E, F^*).$$

Proof: Let $f: E \widehat{\otimes}_{\gamma_2^*} F \to \mathbb{C}$ be a bounded linear form will norm ≤ 1. Clearly f can be identified with a linear operator $u: E \to F^*$ such that

$$\forall \, x \in E \quad \forall \, \xi \in F \qquad f(x \otimes \xi) = \langle \xi, u(x) \rangle.$$

Then (by (5.1)) the inequality

$$(5.2) \qquad \forall \, v \in E \otimes F \qquad |f(v)| \leq \gamma_2^*(v)$$

is equivalent to say that if $(y_i) < (x_j)$ in E we have

$$\sum \langle \xi_i, u(y_i) \rangle \leq \left(\sum \|\xi_i\|^2 \right)^{1/2} \left(\sum \|x_j\|^2 \right)^{1/2}$$

or equivalently (5.2) is equivalent to

$$(y_i) < (x_j) \Rightarrow \left(\sum \|u(y_i)\|^2 \right)^{1/2} \leq \left(\sum \|x_j\|^2 \right)^{1/2}.$$

By Theorem 3.4, this is the same as $\gamma_2(u) \leq 1$.
This gives the announced isometric identity since conversely every $u: E \to F^*$ defines a linear form on $E \otimes F$. $\qquad \square$

Proposition 5.4. *Let S, T be arbitrary sets. Consider v in $\ell_1(T) \otimes \ell_1(S)$ with a finitely supported associated matrix $(v(s, t))$. We have then $\gamma_2^*(v) = \alpha(v)$ where $\alpha(v)$ is defined by*

$$(5.3) \qquad \alpha(v) = \inf \left\{ \left(\sum |\beta_t|^2 \right)^{1/2} \left(\sum |\alpha_s|^2 \right)^{1/2} \|(a(s,t))\|_{B(\ell_2(T), \ell_2(S))} \right\}$$

where the infimum runs over all representations of v of the form

$$v = \sum_{\substack{t \in T \\ s \in S}} \alpha_s a(s, t) \beta_t \; e_t \otimes e_s.$$

Proof: Note that if v is as above we have

$$v = \sum_{s \in S} y_s \otimes \xi_s$$

with $\xi_s = \alpha_s e_s$, $y_s = \sum_t a(s,t) x_t$, $x_t = \beta_t e_t$. If $\|(a(s,t))\| \leq 1$, then $(y_s) < (x_t)$ hence by (5.1) we have $\gamma_2^*(v) \leq \left(\sum |\beta_t|^2\right)^{1/2} \left(\sum |\alpha_s|^2\right)^{1/2}$ so we obtain $\gamma_2^*(v) \leq \alpha(v)$.

The converse follows from Lemma 5.2. Indeed, first note that since v is finitely supported we may reduce to the case when S and T are finite sets. Then assume $\gamma_2^*(v) < 1$ so that there are (y_i), (x_j) in $\ell_1(T)$ and (ξ_i) in $\ell_1(S)$ such that $(y_i) < (x_j)$, $v = \sum y_i \otimes \xi_i$ and $\left(\sum \|x_j\|^2\right)^{1/2} \leq 1$, $\left(\sum \|\xi_i\|^2\right)^{1/2} \leq 1$. Then by Lemma 5.2 applied twice (to $(y_i) < (x_j)$ and to $(\xi_i) < (\xi_i)$) we find on one hand a matrix $B = (b(i,t))$ and scalars β_t with $\sum |\beta_t|^2 \leq 1$ such that $\|B\| \leq 1$ and $y_i = \sum_t b(i,t) \beta_t e_t$ and on the other hand a matrix $C = (c(i,s))$ with $\|C\| \leq 1$ and scalars (α_s) with $\sum |\alpha_s|^2 \leq 1$ such that $\xi_i = \sum_s c(i,s) \alpha_s e_s$.

This yields finally

$$v = \sum y_i \otimes \xi_i = \sum_{s,t} \alpha_s a(s,t) \beta_t \ e_t \otimes e_s$$

where $a(s,t) = \sum_i c(i,s) b(i,t)$, or equivalently $A = {}^t C B$, so that $\|A\| \leq 1$. Thus we obtain $\alpha(v) \leq 1$. This concludes the proof that $\alpha(v) = \gamma_2^*(v)$. $\qquad \square$

Proof of Theorem 5.1. The equivalence (ii) $\Leftrightarrow$ (iii) is easy to check by entirely elementary arguments left to the reader. We now show that (i) is equivalent to (iii). Assume (i). We will show that φ defines a bounded linear form on $\ell_1(T) \hat{\otimes}_{\gamma_2^*} \ell_1(S)$ and hence an operator in $\Gamma_2(\ell_1(T), \ell_\infty(S))$. To check this, by density it suffices to show that for any v in $\ell_1(T) \otimes \ell_1(S)$, of the form $v = \sum_{s,t} v(s,t) e_t \otimes e_s$ with $(v(s,t))$ finitely supported, we have

$$(5.4) \qquad \left| \sum_{s,t} v(s,t) \varphi(s,t) \right| \leq C \gamma_2^*(v).$$

But we observe that the inequality $\left| \sum v(s,t) \varphi(s,t) \right| \leq \alpha(v) \|M_\varphi\|$ is immediate from the definitions, hence the preceding Proposition 5.4 yields (5.4).

The converse is essentially the same argument reversed. If $\gamma_2(u_\varphi) \leq C$, with $u_\varphi \colon \ell_1(T) \to \ell_\infty(S)$ associated to the matrix $(\varphi(s,t))$ then by Proposition 5.4 and Theorem 5.3 again we have

$$\forall \ v \in \ell_1(T) \otimes \ell_1(S) \qquad |\langle u_\varphi, v \rangle| \leq C \alpha(v)$$

which immediately yields by (5.3) that $\|M_\varphi\| \leq C$. This yields (i) $\Leftrightarrow$ (iii).

Finally we check (ii) $\Leftrightarrow$ (iv). Assume (ii). Let $\pi \colon B(\ell_2(T), \ell_2(S)) \to B(\ell_2(T,H), \ell_2(S,H))$ be the mapping $a \to a \otimes I_H$. Clearly π is c.b. and $\|\pi\|_{cb} \leq 1$. (It is easy to reduce to $T = S$ in which case π is a *-representation.) Let $V_1 \colon \ell_2(T) \to \ell_2(T,H)$ and $V_2 \colon \ell_2(S) \to \ell_2(S,H)$ be the operators defined by

$$\forall\, \alpha \in \ell_2(T), \quad \forall\, \beta \in \ell_2(S) \qquad V_1\alpha = (\alpha(t)x_t)_{t\in T} \quad V_2\beta = (\beta(s)y_s)_{s\in S}.$$

Then clearly $\|V_1\|\,\|V_2\| \leq \sup_S \|y_s\| \sup_T \|x_t\| \leq C$ and we have

$$\forall\, a \in B(\ell_2(T), \ell_2(S)) \qquad M_\varphi(a) = V_2^*\pi(a)V_1$$

hence $\|M_\varphi\|_{cb} \leq \|V_2\|\,\|V_1\| \leq C$. This shows that (ii) $\Rightarrow$ (iv). Since (iv) $\Rightarrow$ (i) is trivial, this completes the proof of Theorem 5.1. $\qquad\qquad\square$

We now turn to a theorem of Grothendieck which considerably strengthens Theorem 5.1 at the cost of a multiplicative numerical constant K, the best value of which is called the Grothendieck constant and is denoted by K_G.

Theorem 5.5. *Let S, T be arbitrary sets. Let us denote simply by C the set of all Schur multipliers of the form*

(5.5) $$\varphi(s,t) = \alpha(s)\beta(t)$$

with $\alpha(s)$, $\beta(t)$ scalars such that

$$\sup_{s\in S} |\alpha(s)| \leq 1, \quad \sup_{t\in T} |\beta(t)| \leq 1.$$

Let us denote by $\widetilde{C}$ the closure of $conv(C)$ (the convex hull of C) for the topology of pointwise convergence on $S \times T$. Then there is an absolute constant K such that

(5.6) $$\widetilde{C} \subset \{\varphi \mid \|M_\varphi\| \leq 1\} \subset K\widetilde{C}.$$

The preceding result reduces many questions on Schur multipliers to the much simpler multipliers of the form (5.5).

We will give Krivine's proof of Grothendieck's theorem, which yields the best known estimate of K_G, i.e. of the smallest constant K for which (5.6) holds. It should be emphasized that this constant depends on the choice of the field of scalars. We will denote by $K_G^{\mathbb{R}}$ and $K_G^{\mathbb{C}}$ the constants in the real and complex case respectively. It is a rather easy exercise to show that $K_G^{\mathbb{C}} \leq K_G^{\mathbb{R}}$, a very easy one to show $K_G^{\mathbb{C}} \leq 2K_G^{\mathbb{R}}$ (and actually it is known that $K_G^{\mathbb{C}} < K_G^{\mathbb{R}}$). Therefore we will give the proof in the real case only.

We will use the following

Lemma 5.6. *Let $(\Omega, \mathcal{A}, P)$ be a probability space and let $G \subset L_2(\Omega, \mathcal{A}, P)$ be a linear subspace formed entirely of Gaussian variables. Then for all g, g' in G with $\|g\|_2 = \|g'\|_2 = 1$ we have*

$$\langle g, g'\rangle = \sin[(\pi/2)\mathbb{E}(\mathrm{sign}(g)\mathrm{sign}(g'))].$$

Proof: Let $\sin\theta_0 = \langle g, g'\rangle$ with $\theta_0 \in [-\pi/2, \pi/2]$. Let us write briefly $s(x) = \mathrm{sign}(x)$. We have

$$\mathbb{E}(s(g)s(g')) = \int\!\!\int s(x)s(x\sin\theta_0 + y\cos\theta_0)e^{-(x^2+y^2)/2}\frac{dx\,dy}{2\pi}.$$

Hence in polar coordinates

$$= \int\int s(\cos\theta)s(\sin(\theta + \theta_0))e^{-r^2/2}rdr\frac{d\theta}{2\pi}$$

$$= \int_0^{2\pi} s(\cos\theta)s(\sin(\theta + \theta_0))\frac{d\theta}{2\pi}$$

$$= \frac{2}{\pi}\theta_0.$$

□

Proof of Theorem 5.5. We will prove only the real case. (The complex version of Lemma 5.6 is more complicated, cf. [H5]) Let $K = \pi(2\operatorname{Log}(1+\sqrt{2}))^{-1}$. This number is chosen so that $K = \pi/2c^2$ with $c = (\operatorname{Log}(1+\sqrt{2}))^{1/2}$ adjusted so that

$$\sinh c^2 = 1.$$

Note that the unit ball $\{\varphi \mid \|M_\varphi\| \leq 1\}$ of the space of Schur multipliers is convex, pointwise closed and obviously contains C. Hence it contains $\widetilde{C}$. The nontrivial inclusion in (5.6) is the converse. Let us denote by S the space of all Schur multipliers $\varphi\colon S \times T \to \mathbb{C}$. We set $\|\varphi\|_S = \|M_\varphi\|$.

By homogeneity, to complete the proof we may as well assume that

(5.7) $$\|\varphi\|_S = \|M_\varphi\| < c.$$

We now use the simple observation that Schur multipliers form a Banach algebra for the pointwise multiplication, so that using a Taylor expansion of $z \to \sin z$ we find

$$\|\sin(c\varphi)\|_S \leq \sinh(c\|\varphi\|_S) < \sinh c^2 = 1.$$

By Theorem 5.1, this means that there are x_t, y_s in some Hilbert space H with $\|x_t\|_H \leq 1$, $\|y_s\|_H \leq 1$ such that

$$\forall t \in T \quad \forall s \in S \quad \sin(c\varphi(s,t)) = \langle x_t, y_s \rangle.$$

Since H is isometric to the closed linear span $G \subset L_2(\Omega, \mathcal{A}, P)$ of a suitable collection of independent standard Gaussian variables, we can assume $H = G$. Moreover, we may as well assume $\|x_t\|_G = \|y_s\|_G = 1$ for all t and s. (Indeed, let x' be a norm one element orthogonal to $\{x_t\} \cup \{y_s\}$, we may replace each x_t by $x_t + (1 - \|x_t\|^2)^{1/2}x'$ and similarly for each y_s.) Now by Lemma 5.6 we have

(5.8) $$c\varphi(s,t) = \frac{\pi}{2}\mathbb{E}[\operatorname{sign}(x_t)\operatorname{sign}(y_s)].$$

(Note: since $|\varphi(s,t)| \leq c < \pi/2c$, equality modulo 2π in (5.8) implies equality.) We now assume that S and T are finite sets. Observe that for any ω in Ω the multiplier

$$(s,t) \longrightarrow \operatorname{sign}(x_t(\omega))\operatorname{sign}(y_s(\omega))$$

is in C, therefore (5.8) implies that $\varphi \in (\pi/2c)\operatorname{conv}(C)$. By homogeneity recalling our initial assumption (5.7), this means that we have proved that if S, T are finite sets

$$\{\varphi \mid \|M_\varphi\| \leq 1\} \subset (\pi/2c^2)\operatorname{conv}(C).$$

This concludes the proof with $K = \pi/2c^2$ in the finite case. Obviously this shows that if φ is finitely supported and $\|M_\varphi\| \leq 1$ then $\varphi \in (\pi/2c^2) \operatorname{conv}(C)$. To obtain the general case, for any finite subset $i \subset S \times T$ of product form (i.e. $i = S' \times T'$ with S', T' finite subsets), we denote by $\varphi_i \colon S \times T \to \mathbb{R}$ the function which coincides with φ on i and vanishes elsewhere. We consider the collection of such subsets i as a net directed by inclusion. Then $\varphi_i \to \varphi$ pointwise, and $\|M_{\varphi_i}\| \leq \|M_\varphi\|$. This shows that $\|M_\varphi\| \leq 1$ implies that φ belongs to the pointwise closure of $(\pi/2c^2) \operatorname{conv}(C)$, or equivalently to $K\widetilde{C}$. $\square$

We now state the classical form of Grothendieck's inequality.

Corollary 5.7. *Let (a_{ij}) be an $n \times n$ matrix of scalars such that*

$$(5.9) \qquad \sup_{\substack{(z_i) \in \mathbf{T}^n \\ (z_i') \in \mathbf{T}^n}} \left| \sum a_{ij} z_i z_j' \right| \leq 1.$$

Then for all n-tuples (x_i) and (y_j) in the unit ball of a Hilbert space we have

$$(5.10) \qquad \left| \sum_{ij} a_{ij} \langle x_j, y_i \rangle \right| \leq K$$

where K is the same constant as in Theorem 5.5.

Proof: By theorem 5.5 and by convexity, (5.9) implies $\left| \sum a_{ij} \varphi(i,j) \right| \leq K_G$ for all φ in the unit ball of the space of Schur multipliers of M_n. But by Theorem 5.1, if $\sup \|x_i\| \leq 1$ and $\sup \|y_i\| \leq 1$ then the multiplier φ defined by

$$\varphi(i,j) = \langle x_j, y_i \rangle$$

is in the latter unit ball. Hence we obtain (5.10). $\square$

Let us denote by the K_n the smallest constant K such that Corollary 5.7 holds for a given (fixed) integer $n \geq 1$. In other words K_n is the (complex) Grothendieck constant restricted to $n \times n$ matrices.

We will need also the following variant: Let $H = \ell_2(I)$ with I an infinite set. We need to consider the standard bilinear form (not the scalar product) defined as follows

$$\forall x, y \in H \qquad \Phi(x,y) = \sum_{i \in I} x_i y_i.$$

So Φ is trivially a symmetric bilinear form of norm 1 on H. We will denote by $\widetilde{K}_n$ the smallest constant $\widetilde{K}$ such that, for any $n \times n$ matrix (a_{ij}) satisfying (5.9), we have for any n-tuple (x_i) in the unit ball of H

$$\left| \sum_{i,j=1}^{n} a_{ij} \Phi(x_i, x_j) \right| \leq \widetilde{K}.$$

We need to observe that

(5.11) $$K_n \leq \widetilde{K}_{2n}.$$

Indeed, given arbitrary n-tuples (x_i), (y_i) we can form the $(2n)$-tuple

$$(X_k) = (\bar{y}_1, \ldots, \bar{y}_n, x_1, \ldots, x_n)$$

(where $y \rightarrow \bar{y}$ denotes the standard antilinear isometry of $H = \ell_2(I)$) and we introduce the $2n \times 2n$ matrix

$$(\tilde{a}_{k\ell}) = \begin{pmatrix} 0 & \vdots & (a_{ij}) \\ \cdots\cdots\cdots\cdots \\ 0 & \vdots & 0 \end{pmatrix}.$$

Then $(\tilde{a}_{k\ell})$ still satisfies the estimate (5.9) and we have

$$\sum_{k,\ell=1}^{2n} \tilde{a}_{k\ell}\Phi(X_\ell, X_k) = \sum_{i,j=1}^{n} a_{ij}\langle x_j, y_i\rangle,$$

so that we easily derive our announced claim (5.11).

The following interesting connection with von Neumann's inequality was observed by Varapoulos [V3].

Theorem 5.8. *Fix an integer $n \geq 1$. For any homogeneous polynomial of degree 2*

(5.12) $$P(z_1, \ldots, z_n) = \sum_{i,j=1}^{n} a_{ij} z_i z_j$$

let $\|P\|_\infty = \sup_{(z_i)\in \mathbf{T}^n} |P(z_1, \ldots, z_n)|$. Let $\widehat{K}_n$ be the smallest constant $\widehat{K}$ such that for any n-tuple $T_1, \ldots, T_n$ of mutually commuting contractions on an arbitrary Hilbert space, we have for any such P

$$\|P(T_1, \ldots, T_n)\| \leq \widehat{K}\|P\|_\infty.$$

Then

(5.13) $$\widetilde{K}_n \leq \widehat{K}_n$$

hence by (5.11)

(5.14) $$K_n \leq \widehat{K}_{2n}$$

so that

(5.15) $$K_G = \sup_{n\geq 1} K_n \leq \sup_{n\geq 1} \widehat{K}_n.$$

Remark. It is well known that $K_G > 1$ in either the real or complex case. Actually we will show $K_G \geq \frac{4}{\pi}$ in the complex case and $K_G \geq \frac{\pi}{2}$ in the real one. This minoration goes back to Grothendieck [G]. (For better bounds, see [Kr1,

H5, Kö].) First observe that for any measure space (Ω, μ) and for any linear operator $u\colon L_\infty(\Omega, \mu) \to L_1(\Omega, \mu)$ we have for all finite sequences (x_i) and (y_i) in $L_\infty(\Omega, \mu)$

$$(5.16) \quad \left| \sum \langle u(x_i), y_i \rangle \right| \leq K_G \|u\| \left\| \left(\sum |x_i|^2 \right)^{1/2} \right\|_\infty \left\| \left(\sum |y_i|^2 \right)^{1/2} \right\|_\infty.$$

Indeed, this is the same as Corollary 5.7 except that we have passed from a discrete measure space (namely $\Omega = \{1, \ldots, n\}$ with counting measure) to a general one. The passage from the discrete to the continuous is a routine technique which we leave to the reader.

Consider then a probability space (Ω, P) and a sequence of independent complex (resp. real) valued random variables (g_i) each with a standard Gaussian distribution, so that $\mathbf{E}(g_i) = 0$ and $\mathbf{E}|g_i|^2 = 1$. For simplicity, we will write L_p instead of $L_p(\Omega, P)$, and $\| \ \|_p$ for the corresponding norm $(1 \leq p \leq \infty)$. Then let $Q\colon L_2 \to L_2$ be the orthogonal projection onto the closed span of the sequence (g_i). We define then

$$u = JQJ^*\colon L_\infty \to L_1$$

where $J\colon L_2 \to L_1$ is the natural inclusion map. Then we have of course $\|u\| \leq 1$, but this can be improved: we claim that actually

$$(5.17) \quad \|u\| \leq (\|g_1\|_1)^2 = \begin{cases} \dfrac{\pi}{4} & \text{in the complex case,} \\[2mm] \dfrac{2}{\pi} & \text{in the real case.} \end{cases}$$

Indeed, for any $x = \sum \alpha_i g_i$ in $\mathrm{span}(g_i)$, the distribution of x is Gaussian with mean zero and variance $\sum |\alpha_i|^2$, so that

$$\|x\|_1 = \|g_1\|_1 \left(\sum |\alpha_i|^2 \right)^{1/2}$$

which implies

$$\|JQ\colon L_2 \to L_1\| \leq \|g_1\|_1.$$

Taking adjoints, we find

$$\|QJ^*\colon L_\infty \to L_2\| \leq \|g_1\|_1$$

whence

$$\|u\| = \|JQJ^*\| \leq \|g_1\|_1^2.$$

Finally, the numerical values in (5.17) follow from an explicit calculation. We find

$$\|g_1\|_1 = \int_{\mathbb{R}} |t| e^{-\frac{t^2}{2}} \frac{dt}{\sqrt{2\pi}} = \sqrt{\frac{2}{\pi}} \quad \text{in the real case}$$

and

$$\|g_1\|_1 = \int\int_{\mathbb{R}^2} \sqrt{x^2 + y^2}\, e^{-\frac{x^2 + y^2}{2}} \frac{dx\,dy}{2\pi} = \frac{\pi}{4} \quad \text{in the complex case.}$$

But now let us define

$$x_i = y_i = g_i \left(\sum_1^n |g_i|^2 \right)^{-1/2}.$$

For simplicity, let $S_n = \left(\sum_1^n |g_i|^2 \right)^{1/2}$.
We have $\langle x_i, g_j \rangle = 0 \; \forall i \neq j$ hence

$$u(x_i) = \beta_i g_i$$

with $\beta_i = \langle x_i, g_i \rangle = \mathbf{E}(|g_i|^2 S_n^{-1})$. Note that by the distributional invariance under permutations of $(g_1, \ldots, g_n)$ we have $\beta_1 = \cdots = \beta_n = (1/n) \sum_1^n \beta_i = (1/n)\mathbf{E}(\sum_1^n |g_i|^2 S_n^{-1}) = (1/n)\mathbf{E}(S_n)$. Moreover

$$\langle u(x_i), y_i \rangle = \beta_i \langle g_i, y_i \rangle = \beta_i^2$$

so that

$$\sum_1^n \langle u(x_i), y_i \rangle = \sum_1^n \beta_i^2 = n^{-1}(\mathbf{E}(S_n))^2 = [n^{-1/2}\mathbf{E}(S_n)]^2.$$

But by the strong (resp. weak) law of large numbers we have

$$n^{-1/2}S_n \to 1 \quad \text{a.s. (resp. } n^{-1/2}\mathbf{E}(S_n) \to 1)$$

hence we obtain from (5.16) and (5.17)

$$1 \leq K_G \|u\| \leq K_G (\|g_1\|_1)^2$$

and we conclude as announced that $K_G \geq 4/\pi$ in the complex case (and $\geq \pi/2$ in the real one). Whence:

Corollary 5.9. $\sup_{n \geq 1} \widehat{K}_n \geq 4/\pi > 1$. *In particular for some* n, *the von Neumann inequality fails for* n *mutually commuting contractions and for a homogeneous polynomial of degree equal to 2.*

Proof of Theorem 5.8. We use the same notation as above. Now consider $x_1, \ldots, x_n$ in the unit ball of H and let (a_{ij}) be an $n \times n$ matrix satisfying (5.9). Then the polynomial P defined by (5.12) satisfies (a fortiori) $\|P\|_\infty \leq 1$. We now consider the (Hilbertian) direct sum

$$\mathcal{H} = \mathbb{C} \oplus H \oplus \mathbb{C}$$

and we denote $e = 1 \oplus 0 \oplus 0$ and $f = 0 \oplus 0 \oplus 1$. We will consider H as embedded into $\mathcal{H}$. We define an operator $T_i \in B(\mathcal{H})$ as follows

$$T_i e = x_i$$
$$\forall y \in H \qquad T_i y = \Phi(x_i, y)f$$
$$T_i f = 0.$$

Then (using the symmetry of Φ) it is easy to check that the resulting linear operators (T_i) mutually commute on $\mathcal{H}$ and that $\|T_i\|_{B(\mathcal{H})} \leq \|x_i\|_H \leq 1$. Moreover, we have $P(T_1, \ldots, T_n) = \sum a_{ij} T_i T_j$ hence

$$\langle P(T_1, \ldots, T_n)e, f \rangle = \sum a_{ij} \Phi(x_i, x_j).$$

Hence we find

$$\left| \sum a_{ij} \Phi(x_i, x_j) \right| \leq \|P(T_1, \ldots, T_n)\|$$
$$\leq \widehat{K}_n \|P\|_\infty \leq \widehat{K}_n.$$

This shows (5.13). Now, recalling (5.11), (5.14) and (5.15) are clear.   □

Remark. The exact values of the first integer n for which $\widehat{K}_n > 1$ and of the first one for which $K_n > 1$ seem unclear, as far as I know. Recently, Quanhua Xu observed that $K_2 = 1$ (personal communication). I have been informed by Allan Sinclair that this was already known to A. M. Davie. The reader should be warned that there is a different notion of n-dimensional Grothendieck constant in the literature, where one considers in Corollary 5.7 matrices with arbitrary size but vectors in an n-dimensional Hilbert space. Let us denote this constant by $K_G^{\mathbb{C}}(n)$, as in [Kö]. Then, by results due to Davie and König (see [Kö]), we have $K_G^{\mathbb{C}}(2) > 1$. Then, it follows from the argument for Corollary 5.9 that, for some integer N, there are N mutually commuting contractions *on a 6-dimensional Hilbert space H* failing von Neumann's inequality with respect to a homogeneous polynomial of degree 2. It is tempting to try to determine the smallest possible value d of the dimension of H, but we have not gone into this. (Note that $d > 2$ by [Dr].)

The reader in need of "exercises" can use for that purpose the following known results that we state without proof (but with some indication of proof). They are all concerned with generalizations to the L_p-case of statements proved above in the Hilbert space case.

By an L_p-space, we mean a Banach space of the form $L_p(\Omega, \mathcal{A}, \mu)$ for some measure space $(\Omega, \mathcal{A}, \mu)$. Of course, since the measure μ is entirely unrestricted, this includes spaces such as $\ell_p(I)$ for any index set I (in that case μ is the measure on I which gives mass one to each point). We will say that an operator $u\colon X \to Y$ factors through L_p (resp. through a subspace of L_p, resp. through a quotient of L_p, resp. through a subspace of a quotient of L_p) if there is a Banach space Z isometric to an L_p-space (resp. to a subspace of an L_p-space, resp. to a quotient of an L_p-space, resp. to a subspace of a quotient of an L_p-space) and operators $B\colon X \to Z$ and $A\colon Z \to Y$ such that $u = AB$.

We denote by $\Gamma_p(X,Y)$ (resp. $\Gamma_{S_p}(X,Y)$, resp. $\Gamma_{Q_p}(X,Y)$, resp. $\Gamma_{SQ_p}(X,Y)$) the space of all such operators and we let $\gamma_p(u)$ (resp. $\gamma_{S_p}(u)$, resp. $\gamma_{Q_p}(u)$, resp. $\gamma_{SQ_p}(u)$) be the infimum of $\|A\|\,\|B\|$ over all such factorizations. It can be proved (actually this follows from Theorem 5.10 below) that $\Gamma_p(X,Y)$ (resp. $\Gamma_{S_p}(X,Y)$, resp. $\Gamma_{Q_p}(X,Y)$, resp. $\Gamma_{SQ_p}(X,Y)$) equipped with this norm is a Banach space. The reader should take note that the class of subspaces of quotients of L_p coincides with the class of quotients of subspaces of L_p, so that there is no need to go further and consider iterated spaces like Γ_{QS_p}, Γ_{QSQ_p} or Γ_{SQS_p} and so on. They are all the same as Γ_{SQ_p}.

Let $(y_i), (x_j)$ be finite sequences in a Banach space X. Let $1 \le p < \infty$. By the notation $(y_i) <_p (x_j)$ we will mean that

$$\forall \xi \in X^* \qquad \sum |\xi(y_i)|^p \le \sum |\xi(x_j)|^p.$$

(So if $p = 2$, this was denoted previously by $(y_i) < (x_j)$.).

Theorem 5.10. *Let $u\colon X \to Y$ be an operator and let C be a constant. The following are equivalent*

(i) *For all finite sequences $(y_i), (x_j)$ in X such that $(y_i) <_p (x_j)$, we have*

$$\sum \|u(y_i)\|^p \le C^p \sum \|x_j\|^p.$$

(ii) *There are a measure space (Ω, μ), a subspace $Z \subset L_p(\mu)$ and operators $B\colon X \to Z$ and $A\colon Z \to Y$ such that $u = AB$ and $\|A\|\,\|B\| \le C$.*

This result was first proved in [LiP]. It can be proved exactly as Theorem 3.4 above in the case $p = 2$, but using Kakutani's well known characterization of L_p-spaces as Banach lattices such that $\|x + y\| = (\|x\|^p + \|y\|^p)^{1/p}$ whenever $|x| \wedge |y| = 0$. Using Theorem 5.10 instead of Theorem 3.4 and an obvious extension of Lemma 5.2, we immediately obtain

Theorem 5.11. *Let $\varphi\colon S \times T \to \mathbb{C}$ be a function and let $C \ge 0$ be a constant. The following are equivalent if $1 \le p < \infty$ and $\frac{1}{p} + \frac{1}{p'} = 1$.*

(i) *φ is a bounded Schur multiplier on $B(\ell_p(T), \ell_p(S))$ with norm $\le C$.*

(ii) *There is a measure space (Ω, μ) and elements $(x_t)_{t \in T}$ in $L_p(\mu)$ and $(y_s)_{s \in S}$ in $L_{p'}(\mu)$*

$$\forall\, s \in S \quad \forall\, t \in T \qquad \varphi(s,t) = \langle x_t, y_s \rangle \quad and \quad \sup_{s \in S} \|y_s\|_{p'} \sup_{t \in T} \|x_t\|_p \le C.$$

(iii) *The operator $u_\varphi \colon \ell_1(T) \to \ell_\infty(S)$ which admits $(\varphi(s,t))$ as its matrix is in the space $\Gamma_p(\ell_1(T), \ell_\infty(S))$ and $\gamma_p(u_\varphi) \le C$.*

Remark. See Corollary 8.2 below for the "completely bounded" version of (i).
Remark. Note that the case $p = 1$ is trivial, in that case the above properties are all equivalent to $\sup |\varphi(s,t)| \le C$. Moreover, we have $\|u_\varphi\| = \sup_{S \times T} |\varphi(s,t)|$.
Remark. Let $u \colon X \to Y$ be an operator between Banach spaces. Assume X arbitrary and $Y = \ell_\infty(S)$ (resp. $X = \ell_1(T)$ and Y arbitrary), then

$$\Gamma_{S_p}(X,Y) = \Gamma_p(X,Y) \qquad (\text{resp.} \quad \Gamma_{Q_p}(X,Y) = \Gamma_p(X,Y))$$

and

$$\gamma_{S_p}(u) = \gamma_p(u) \qquad (\text{resp.} \quad \gamma_{Q_p}(u) = \gamma_p(u)).$$

Indeed, this is a simple consequence of the lifting property of $\ell_1(T)$ and of the extension property of $\ell_\infty(S)$. More generally, using both properties (lifting and extension), for any $u \colon \ell_1(T) \to \ell_\infty(S)$ we have

$$\gamma_p(u) = \gamma_{S_p}(u) = \gamma_{Q_p}(u) = \gamma_{SQ_p}(u).$$

Notes and Remarks on Chapter 5

Although it has been rediscovered by many subsequent authors, the fundamental result on Schur multipliers which we state as Theorem 5.1 essentially goes back to Grothendieck. Indeed, in [G, Proposition 7, p. 68], Grothendieck states a result which clearly implies the equivalence of the first three conditions in Theorem 5.1. Note however that his formulation does not give precisely the equality of the corresponding three constants but only their equivalence. Grothendieck's methods recurrently appeal to decompositions into the positive and negative parts of a quadratic form and sometimes this leads to an extra factor 2 or 4. For instance, at the end of the "Résumé" (this is how Banach space theorists refer to [G]), Grothendieck asks a question which, with the notation of Chapter 5, can be rephrased as: what is the best constant C such that $\gamma_2 \leq C\gamma_2^*$? He only obtained $C \leq 2$ by his methods. Later on, Pietsch ([Pie], see also [DF]) and Kwapień [Kw] developed the duality theory for ideals of operators in a different manner which, in particular, eliminates these extra numerical factors. For instance, it became clear from [Kw] that we have $\gamma_2^* \leq \gamma_2$ so that $C = 1$ is the best constant in that particular question of Grothendieck.

More recently, the equivalence of these three conditions was rediscovered by J. Gilbert [Gi1-2] in connection with the harmonic analysis of Fourier multipliers on function spaces. The same result together with the fourth condition was also found by U. Haagerup [H4], together with various interesting generalizations for operator algebras. Several results on Schur multipliers on various matrix spaces appear in Grahame Bennett's paper [Be].

Lemma 5.2 is a folklore result in Banach space theory, closely related to the Pietsch factorization theorem for 2-absolutely summing operators on the space c_0. The duality Theorem 5.3 appears explicitly in Kwapień's work, but there is an analogous result already present in the "Résumé" [G] (with additional numerical factors). Moreover, in their paper [LiP] (based on [G]), Lindenstrauss and Pełczyński give a criterion (stated above as Theorem 3.4) for an operator to factor through a Hilbert space, and Theorem 5.3 can be viewed as a reformulation of their criterion. Proposition 5.4 is a simple consequence of Theorem 5.3, modulo Lemma 5.2. More recently, we introduced and studied in [P2] a new class of tensor norms, (called the "gamma" norms) which behave like the norms γ_2 and γ_2^*. In fact, in this theory γ_2 (resp. γ_2^*) appears as the smallest (resp. largest) of the gamma norms.

The result stated above as Theorem 5.5 is equivalent to Grothendieck's "fundamental theorem of the metric theory of tensor products", which is the central result of [G]. We should point out that Theorem 5.5, as formulated above, is explicitly mentioned in [G, p. 68] (note that Grothendieck considers Schur multipliers on the predual $\ell_2(S) \widehat{\otimes} \ell_2(T)$ instead of $B(\ell_2(S), \ell_2(T))$, but they are clearly the same).

The proof of Theorem 5.5 included in the text is due to Krivine [Kr2] (see also [Kr1]). It gives the best known estimate for the constant K_G in the real case (and Krivine conjectures that it is sharp). See [H5] for the complex case. We refer the interested reader to the discussion in [P1, Chapter 5] for more information. Corollary 5.7 is the best known form of Grothendieck's inequality as emphasized in [LiP].

Theorem 5.8 and Corollary 5.9 are due to Varopoulos [V3].

The characterization of operators in $\Gamma_{S_p}(X, Y)$, stated above as Theorem 5.10, is due to Lindenstrauss and Pełczyński [LiP], but this direction was greatly expanded by the work of Kwapień [Kw] which contains characterizations of operators in the classes $\Gamma_p(X, Y)$ and $\Gamma_{SQ_p}(X, Y)$. See the recent books [DF] and [DJT] for more information on these classes and their relationship to absolutely summing operators.

The study of the Schur multipliers for the spaces $B(L_p, L_p)$ was apparently initiated by Herz [Her1] (see also [Her2] for an early study of the class of subspaces of quotient of L_p, under a different name). Once Kwapień had extended the factorization theorems to the L_p-case, it is probably fair to say that it was not too difficult to also extend the theory of Schur multipliers, but Kwapień's results were probably not too well known outside Banach space theory. Thus, the result we state as Theorem 5.11 was rediscovered by many authors. It seems to go back to Gilbert's work [Gi2]. See also Fendler's thesis [Fe2], [B1] and [CoF] for more results.

6. Hankelian Schur multipliers. Herz-Schur multipliers

Summary: In this short chapter, we discuss Schur multipliers restricted to various subspaces $E \subset B(H)$. We first discuss the case when $H = \ell_2$ and E is the sub-class of all Hankel matrices. We show that the Schur multipliers which are completely bounded maps from E to E are closely related to the Fourier multipliers on the Hardy space H_1. Analogously, when $H = \ell_2(G)$ and E is the reduced C^*-algebra $C^*_\lambda(G)$, then the Schur multipliers which are completely bounded maps from E to E are identical to the completely bounded multipliers of $C^*_\lambda(G)$ or equivalently to the (so-called) Herz-Schur multipliers of G.

Let $Hank \subset B(\ell_2)$ be the subspace of $B(\ell_2)$ formed of all the Hankelian matrices, i.e. matrices of the form $a = (f(i+j))_{i,j \in \mathbb{N}}$ for some function $f \colon \mathbb{N} \to \mathbb{C}$. Let us denote

$$L_n = \sum_{i+j=n} e_i \otimes e_j.$$

Then $Hank$ is the weak-$*$ closed linear span of $\{L_n | n \geq 0\}$.
By a well known theorem of Nehari (see e.g. [Ni]) all such matrices a can be written as

$$a = (\widehat{\varphi}(-(i+j)))$$

for some φ in $L_\infty(\mathbf{T})$ (the minus sign is here for convenience) and moreover

(6.1) $$\|a\|_{B(\ell_2)} = \inf\{\|\varphi\|_\infty\}$$

where the infimum runs over all possible representations of a.
In other words, we have a map

$$Q \colon L_\infty(\mathbf{T}) \to B(\ell_2)$$

defined by

$$Q(\varphi)_{ij} = \widehat{\varphi}(-(i+j))$$

and this map is a (metric) surjection of $L_\infty(\mathbf{T})$ onto $Hank$. Since

$$\ker(Q) = H^0_\infty = \{f \in L_\infty | \hat{f}(n) = 0 \ \forall n \leq 0\},$$

we have an isometric identification

$$Hank = L_\infty / H^0_\infty.$$

Let us observe that Q is c.b. with $\|Q\|_{cb} \leq 1$. It can be proved using the fact that compressions of c.b. maps are again c.b. maps. Indeed, let $\pi\colon L_\infty(\mathbf{T}) \to B(L_2(\mathbf{T}))$ be the representation which maps φ to the operator of multiplication by φ. If we identify H_2 with ℓ_2, we may view a Hankel operator in $Hank$ as a bounded bilinear form on $H_2 \times H_2$, or equivalently as an operator from H_2 to $H_2^* = \bar{H}_2$. With this identification we have

$$Q(\varphi) = P_{\bar{H}_2} \pi(\varphi)_{|H_2}.$$

Since π is a *-representation, this implies $\|Q\|_{cb} \leq \|\pi\|_{cb} \leq 1$.

Theorem 6.1. *For a sequence of scalars $(m(n))_{n \geq 0}$, the following are equivalent.*

(i) *The linear operator θ, unambiguously defined on the linear span of $\{L_n\}$ by $\theta(L_n) = m(n)L_n$ extends by weak-* density to a c.b. map from $Hank$ into itself with c.b. norm ≤ 1.*
(ii) *The operator θ extends to a Schur multiplier on $B(\ell_2)$ with norm ≤ 1.*
(iii) *The sequence $(m(i+j))_{i,j \geq 0}$ defines a bounded Schur multiplier $T\colon B(\ell_2) \to B(\ell_2)$ with norm ≤ 1.*

Proof: Clearly (iii) is but a rephrasing of (ii), and (iii) $\Rightarrow$ (i) is clear by Theorem 5.1. Note that Schur multipliers of the form $(m(i+j))$ (i.e. of Hankelian form) leave the subspace $Hank$ invariant. Hence it suffices to show (i) $\Rightarrow$ (iii). Assume (i). By the preceding remark the map $\theta Q\colon L_\infty(\mathbf{T}) \to B(\ell_2)$ is c.b. with c.b. norm ≤ 1. By the factorization of c.b. maps (Theorem 3.6) there is a representation $\pi\colon L_\infty(\mathbf{T}) \to B(\widehat{H})$ and operators $V_1, V_2\colon \ell_2 \to \widehat{H}$ with $\|V_1\|\,\|V_2\| \leq 1$ such that

$$\forall\, n \geq 0 \qquad m(n)L_n = V_2^* \pi(z^n) V_1.$$

Hence for all $i \geq 0$, $j \geq 0$

$$\begin{aligned} m(i+j) &= \langle m(i+j)L_{i+j}, e_i \otimes e_j \rangle = \langle \pi(z^{i+j})V_1 e_i, V_2 e_j \rangle \\ &= \langle \pi(z^i)V_1 e_i, \pi(z^j)^* V_2 e_j \rangle. \end{aligned}$$

Hence, if we let $x_i = \pi(z^i)V_1 e_i$ and $y_j = \pi(z^j)^* V_2 e_j$, we find

$$\forall\, i, j \geq 0 \qquad m(i+j) = \langle x_i, y_j \rangle$$

and $\sup_i \|x_i\| \cdot \sup_j \|y_j\| \leq \|V_1\|\,\|V_2\| \leq 1$. By Theorem 5.1, if $\varphi(i,j) = m(i+j)$ we have $\|M_\varphi\| \leq 1$ whence (iii).   □

Let S_1 be the space of all the trace class operators on ℓ_2, with its usual norm. Recall $S_1^* = B(\ell_2)$. We denote by $H_1(S_1)$ the space of all H_1-functions with values in S_1, equipped with the natural norm. Similarly, we denote by S_1^n and $H_1(S_1^n)$ (and by $L_\infty(M_n)$, $H_\infty^0(M_n)$,...) the analogous spaces relative to $n \times n$ matrices. Note the isometric identification $H_1(S_1^n)^* = L_\infty(M_n)/H_\infty^0(M_n)$. By a simple dualization (using the vectorial Nehari theorem) one obtains the following fact, which we state for emphasis

Theorem 6.2. *The three conditions in the preceding Theorem are equivalent to the following*

(iii)' *There are sequences (x_j) and (y_i) in the unit ball of Hilbert space such that,*

$$\forall i, j \geq 0 \quad m(i+j) = \langle x_j, y_i \rangle.$$

(iv) *The multiplier $\sum_0^\infty x_n e^{int} \to \sum_0^\infty x_n m(n) e^{int}$ is bounded on $H_1(S_1)$ with norm ≤ 1.*

Proof: Note that (iii)' is but a reformulation of (iii), taking Theorem 5.1 into account.

Let us denote by $T_*: H_1 \to H_1$ the operator associated to the multiplier m, when it makes sense. Similarly we denote by $T_* \otimes I_{S_1}$ (resp. $T_* \otimes I_{S_1^n}$) the corresponding operator on $H_1(S_1)$ (resp. $H_1(S_1^n)$). Note that we have obviously

(6.2) $$\|T_* \otimes I_{S_1}\|_{H_1(S_1) \to H_1(S_1)} = \sup_n \|T_* \otimes I_{S_1^n}\|_{H_1(S_1^n) \to H_1(S_1^n)}.$$

A first proof can be given as follows (see the next remark for an alternate one). Note that by the vectorial Nehari theorem ([Sa or Pag], see also [Ni]) we have the following isometric identity, which is a matrix valued version of (6.1)

(6.3) $$M_n(Hank) = L_\infty(M_n)/H_\infty^0(M_n).$$

By duality, this implies

$$\|T_* \otimes I_{S_1^n}\|_{H_1(S_1^n) \to H_1(S_1^n)} = \|I_{M_n} \otimes T\|_{M_n(Hank) \to M_n(Hank)},$$

hence by (6.2)

$$\|T_* \otimes I_{S_1}\|_{H_1(S_1) \to H_1(S_1)} = \|T\|_{cb},$$

whence the equivalence of (i) and (iv).   □

Remark. Actually, it is rather easy to prove directly that (iv) implies that $(m(i+j))$ is a bounded multiplier of Schur type on S_1, hence by duality that it is bounded on $B(\ell_2)$. Indeed, assume (iv) and let x be an arbitrary element of S_1. Consider the analytic S_1-valued function f defined by setting $f(z) = D_z x D_z$ where D_z denotes the diagonal operator on ℓ_2 with diagonal coefficients $(1, z, z^2, z^3, ...)$. Clearly $\|f\|_{H_1(S_1)} = \|x\|_{S_1}$ and $\|[T_* \otimes I_{S_1}](f)\|_{H_1(S_1)} = \|T_* x\|_{S_1}$. Thus if (iv) holds, T_* is a contraction on S_1, hence by transposition (iii) holds.

Remark 6.3. Recently, Effros-Ruan and Blecher-Paulsen developed a new duality theory of "operator spaces" ([ER, BP2]). In this theory, any L_1 space (even any non-commutative one) can be embedded isometrically in a natural way into $B(H)$. A fortiori, the same is true of any subspace of L_1, such as H_1 for instance. Thus, it now makes sense to say that a linear map $u : L_1 \to L_1$ (or a map $u : H_1 \to H_1$) is completely bounded. According to this theory, a map $u : H_1 \to H_1$ is c.b. iff the operator $u \otimes I_{S_1}$ is bounded on $H_1(S_1)$ and we have

$$\|u\|_{cb} = \|u \otimes I_{S_1}\|_{H_1(S_1) \to H_1(S_1)}.$$

Therefore, Theorem 6.2 can be viewed as a characterization of the completely bounded Fourier multipliers on H_1. Similar ideas can be developed for L_p-spaces (including non-commutative ones). See [P7].

In the rest of this chapter, we replace the semi-group $\mathbb{N}$ by a discrete group G and study the analogous notion of Hankelian Schur multipliers. In the Harmonic Analysis literature, the corresponding multipliers are sometimes called Herz-Schur multipliers. They were considered by Herz [Her1] (in a dual framework, as multipliers on the Fourier algebra $A(G)$) before the notion of complete boundedness surfaced. The next result from [BF1] (see also [H3]) clarifies the relationship between the various kinds of multipliers. It is analogous to the preceding Theorem. Note however that since we now deal with a group (instead of a semi-group), there is no difference between the Toeplitz and the Hankel case.

Theorem 6.4. *Let G be a discrete group. Consider a function $\varphi : G \to \mathbb{C}$. We define complex functions φ_1, φ_2 and φ_3 on $G \times G$ by setting*

$$\forall\,(s,t) \in G \times G \quad \varphi_1(s,t) = \varphi(st^{-1}), \quad \varphi_2(s,t) = \varphi(s^{-1}t), \quad \varphi_3(s,t) = \varphi(st).$$

We consider the corresponding Schur multipliers $M_{\varphi_1}, M_{\varphi_2}$ and M_{φ_3} acting on $B(\ell_2(G))$. Then the following are equivalent

(i) *The linear map defined on the linear span of $\{\lambda(t) \mid t \in G\}$ by $T_\varphi(\lambda(t)) = \varphi(t)\lambda(t)$ extends by norm density to a c.b. mapping*

$$T_\varphi :\; C_\lambda^*(G) \to C_\lambda^*(G)$$

with $\|T_\varphi\|_{cb} \leq 1$.

(ii) *The same linear map extends to a Schur multiplier on $B(\ell_2(G))$ with norm ≤ 1.*

(iii) *The function φ_1 defines a bounded Schur multiplier on $B(\ell_2(G))$ with norm ≤ 1.*

Moreover, M_{φ_1} is bounded iff M_{φ_2} (resp. M_{φ_3}) is bounded, and we have

$$\|M_{\varphi_1}\| = \|M_{\varphi_1}\|_{cb} = \|M_{\varphi_2}\| = \|M_{\varphi_2}\|_{cb} = \|M_{\varphi_3}\| = \|M_{\varphi_3}\|_{cb}.$$

Proof: Recall that by Theorem 5.1 the norm and the cb-norm of a Schur multiplier on $B(H)$ are equal, so the last part is immediate (note that if $\psi(s,t)$ is a bounded Schur multiplier on $S \times T$ then, for any bijections $f : S \to S$ and $g : T \to T$, the composition $\psi(f(s), g(t))$ also is a bounded Schur multiplier with the same norm). Note that M_{φ_1} leaves $C_\lambda^*(G)$ invariant and its restriction to $C_\lambda^*(G)$ coincides with T_φ. Hence we have

$$\|T_\varphi\|_{cb} \leq \|M_{\varphi_1}\|_{cb} = \|M_{\varphi_1}\|.$$

This shows (iii) $\Rightarrow$ (i). Conversely, if $\|T_\varphi\|_{cb} \leq 1$ then by Theorem 3.6, there is a Hilbert space H, a representation $\pi : B(\ell_2(G)) \to B(H)$ and operators V_1 and V_2 from $\ell_2(G)$ into H with $\|V_1\| \leq 1, \|V_2\| \leq 1$ such that $\forall\, a \in C_\lambda^*(G) \quad T_\varphi(a) = V_2^* \pi(a) V_1$.

In particular we have $\varphi(x)\lambda(x) = T_\varphi(\lambda(x)) = V_2^* \pi(\lambda(x))V_1$, which implies

$$\forall \, s,t \in G \quad \varphi(st^{-1}) = \; < \delta_s, T_\varphi(\lambda(st^{-1}))\delta_t >$$
$$= \; < \pi(\lambda(s))^* V_2 \delta_s, \pi(\lambda(t^{-1})) V_1 \delta_t >$$

This shows that φ_1 satisfies (ii) in Theorem 5.1, hence $\|M_{\varphi_1}\| \leq 1$. Thus we obtain (i) $\Rightarrow$ (iii). The equivalence between (ii) and (iii) is obvious. $\Box$

Notes and Remarks on Chapter 6

Theorem 6.1 is a simple observation which apparently has not been recorded yet (but others might have already observed this). Similarly, Theorem 6.2 seems new. Note that, from a Fourier analyst viewpoint, Theorem 6.2 is somewhat striking: indeed, since the multipliers of H_1 apparently do not admit any "nice" description, it is surprising that the "completely bounded" ones do.

For a Hilbert space H, let us denote by $S_p(H)$ the Schatten class of all compact operators $x\colon H \to H$ such that $tr|x|^p < \infty$, equipped with the norm $x \to (tr(|x|^p)^{1/p}$. When $H = \ell_2$, we simply write S_p instead of $S_p(H)$.

Following the ideology of the theory of "operator spaces" (see Remark 6.3 and [P7]) we will say that a map $u\colon H_p \to H_p$ (resp. $u\colon L_p \to L_p$) is completely bounded if $u \otimes I_{S_p}\colon H_p(S_p) \to H_p(S_p)$ (resp. $u \otimes I_{S_p}\colon L_p(S_p) \to L_p(S_p)$) is bounded and we set in both cases

$$\|u\|_{cb} = \|u \otimes I_{S_p}\|.$$

Similarly, we will say that a linear map $u\colon S_p \to S_p$ is c.b. if

$$u \otimes I_{S_p}\colon S_p(\ell_2 \otimes_2 \ell_2) \to S_p(\ell_2 \otimes_2 \ell_2)$$

is bounded and we set again

$$\|u\|_{cb} = \|u \otimes I_{S_p}\|.$$

This terminology is justified by the existence of certain isometric embeddings $L_p \subset B(H)$ and $S_p \subset B(H)$ (which are "natural" from the viewpoint of the theory of operator spaces) relative to which the c.b. norm of u is exactly the same as above (see [P7]).

With this terminology, the result of [LuP] says that there are bounded multipliers of H_1 which are not c.b. Recall that if $1 < p \neq 2 < \infty$, the fundamental system $(e_i \otimes e_j)$ spanning S_p is *not* unconditional. However, if we denote by $((\xi_i)_{i \in \mathbb{N}}, (\eta_i)_{i \in \mathbb{N}})$ a generic point in $\mathbf{T}^{\mathbb{N}} \times \mathbf{T}^{\mathbb{N}}$, then the system of functions $\{\xi_i \eta_j \mid i, j \in \mathbb{N}\}$ is unconditional in $L_p(\mathbf{T}^{\mathbb{N}} \times \mathbf{T}^{\mathbb{N}})$ (relative to Haar measure). Since each multiplier of the form $(\xi_i \eta_j)$ clearly is a norm one Schur multiplier from S_p to itself, it is not too difficult to deduce from the preceding facts that there are bounded Fourier multipliers of $L_p(\mathbf{T}^{\mathbb{N}} \times \mathbf{T}^{\mathbb{N}})$ (or actually of $L_p(\mathbf{T})$ itself using lacunary sequences) which are not completely bounded if $1 < p \neq 2 < \infty$. However, at the time of this writing we could not prove the following.

Conjecture. For $1 < p \neq 2 < \infty$, there are bounded Schur multipliers of S_p which are not completely bounded.

Also, it is a well known open problem to describe all the Schur multipliers of S_p. This might be as hopeless as trying to describe all bounded Fourier multipliers of $L_p(\mathbf{T})$. Nonetheless, there might very well exist a simple description of all the *completely bounded* Schur multipliers of S_p.

Theorem 6.4 (i) is due to Bożejko and Fendler [BF1]. I noticed the above simple proof, but Haagerup informed me that this proof had already been found by P. Jolissaint [Jo]. The second assertion in Theorem 6.4 is entirely elementary.

For various results on Herz-Schur multipliers, we refer the reader to the important papers [DCH] and [CoH].

In the following comments, we try to clarify the interrelations between the various notions considered in the previous chapters.

In [Pe1], Peller introduces for each $c > 1$ the norm $||| \cdot |||_c$ defined as follows: for any analytic polynomial P we set

$$|||P|||_c = \sup\{\|P(T)\|\}$$

where the supremum runs over all power bounded operators $T \colon H \to H$ with $\sup_{n \geq 1} \|T^n\| \leq c$. (The same norm is studied in [B2].) By Proposition 4.11 for $n = 1$, we have

$$|||P|||_c \leq c^2 \|P\|_{H_1 * L_\infty}$$

where we have set

$$\|P\|_{H_1 * L_\infty} = \inf\left\{ \sum_1^\infty \|F_k\|_{H_1} \|\varphi_k\|_{L_\infty} \right\}$$

with the infimum running over all possible representations of P of the form

$$P = \sum_1^\infty F_k * \varphi_k$$

(convolution is meant here on the circle as defined before Proposition 4.16).

In the last section of [Pe1], Peller proposes a list of 10 open problems which aim at finding a functional analytic equivalent of the norm $||| \; |||_c$. For instance, he noted that (by an old idea of Herz) the above norm $\| \; \|_{H_1 * L_\infty}$ is a Banach algebra norm for the pointwise product, *i.e.* there is a constant C such that $\|PQ\|_{H_1 * L_\infty} \leq C\|P\|_{H_1 * L_\infty}\|Q\|_{H_1 * L_\infty}$ for all polynomials P and Q (actually $C = 1$ in this case). He also considered the following norm defined on polynomials $P = \sum a_n z^n$ by

$$\|P\|_{\mathcal{L}} = \sup\{|\sum a_n m(n)|\}$$

where the supremum runs over all the multipliers $(m(n))$ such that the sequence $(m(i+j))_{i,j \geq 0}$ defines a bounded Schur multiplier $T \colon B(\ell_2) \to B(\ell_2)$ with norm ≤ 1, as in Theorem 6.1.

There is another description of this norm: for F in $H_1(S_1)$ and φ in $L_\infty(S_\infty)$, let us denote by $F*\varphi$ the scalar valued function defined on $\mathbb{T}$ by

$$F*\varphi(e^{it}) = \int \langle F(e^{i(t-s)}), \varphi(e^{is}) \rangle \frac{ds}{2\pi}$$

where the duality inside the integral is the one corresponding to $S_1 = (S_\infty)^*$. Then, it is easy to deduce from Theorem 6.2 that

(6.4) $$\|f\|_{\mathcal{L}} = \inf\{\|F\|_{H_1(S_1)}\|\varphi\|_{L_\infty(S_\infty)}\}$$

where the infimum runs over all possible decompositions of f as $f = f*\varphi$.

As Peller observed, we have actually, for all polynomial P,

$$\||P\||_c \le c^2\|P\|_{\mathcal{L}} \le c^2\|P\|_{H^1*L_\infty}.$$

In [Pe1], Peller asked whether $\|\ \|_{\mathcal{L}}$ is a Banach algebra norm, and if yes whether it is an operator algebra norm. The question whether it is a Banach algebra norm has been answered affirmatively by Grahame Bennett in his review of Peller's article ([Math. Reviews Sept. 1983i, 47019]). Note that it can now be verified easily by the same method as Peller for the norm $\|\ \|_{H_1*L_\infty}$, but using the "predual" form (6.4) of Theorem 6.2 to describe the norm $\|\ \|_{\mathcal{L}}$. However, the question whether $\|\ \|_{\mathcal{L}}$ is an operator algebra norm has remained open since. Note that the answer is positive iff, for some $c > 0$, the norms $\||\ \||_c$ and $\|\ \|_{\mathcal{L}}$ are equivalent.

Let X (resp. X_{cb}) denote the completion of the space of all polynomials P equipped with the norm $\|\ \|_{H_1*L_\infty}$ (resp. $\|\ \|_{\mathcal{L}}$). Then it is easy to check that X^* (resp. $(X_{cb})^*$) can be identified isometrically with the space of all bounded (resp. completely bounded) multipliers of H_1. (Use (6.4) for the c.b. case.)

Another question of Peller asked whether the Fourier multipliers of H_1 are included (hence coincide with) the Hankelian Schur multipliers of $B(\ell_2)$. This has been answered negatively by F. Lust-Piquard [LuP] and independently by S. Drury (see [LuP]). This answered negatively questions 2, 3 and 4 from [Pe 1]. Note that, by Theorem 6.2 above, this question boils down to producing a bounded multiplier on H_1 which is not *completely* bounded. Such an example can now also be extracted from section 4 in [HP2], or from [Blo1]. The rest of Peller's questions are apparently still open.

We will only quote here two more of them (namely numbers 7 and 8):

Are all the norms $\||\ \||_c$ equivalent for all $c > 1$?

Is it true that, for any $\varepsilon > 0$, any power bounded operator T on a Hilbert space is similar to an operator $\widetilde{T}$ such that $\sup_{n \ge 1}\|\widetilde{T}^n\| < 1 + \varepsilon$?

Clearly, a positive answer to the second question implies a positive answer to the first one.

It is interesting to compare the preceding discussion with the remarks at the end of chapter 2 on the space $B_c(G)$. Indeed, we can introduce the space $B_c(\mathbb{N})$ as the space of all functions $f: \mathbb{N} \to \mathbb{C}$ for which there is a power bounded operator T on H with $\sup_n \|T^n\| \le c$ and vectors x, y in H such that

$$\forall n \in \mathbb{N} \quad f(n) = \langle T^n x, y \rangle.$$

We equip this space with the norm $\|f\|_{B_c(\mathbb{N})} = \inf\{\|x\| \, \|y\|\}$, where the infimum runs over all possible such representations. This is the Banach space of the "coefficients" of the representations of the semi-group $\mathbb{N}$ which are uniformly bounded by c. Note that for any polynomial $P = \sum a_n z^n$, we have

(6.5) $\qquad \||P\||_c = \sup\{|\sum a_n f(n)| \mid f \in B_c(\mathbb{N}), \, \|f\|_{B_c(\mathbb{N})} \leq 1\}$

Let us denote by $M_{cb}(H_1)$ the space of all completely bounded Fourier multipliers of H_1 in the sense of Remark 6.3 (*i.e.* these are the multipliers satisfying the equivalent conditions of Theorem 6.2). It follows from Theorem 6.2 that $B_c(\mathbb{N}) \subset M_{cb}(H_1)$. Using (6.5), the above question of Peller (whether the norms $\||\ \||_c$ and $\|\ \|_{\mathcal{L}}$ are equivalent) can be reformulated as: Does the converse inclusion holds?

We have previously considered similar questions with a discrete group G in the place of $\mathbb{N}$. Let us denote by $C_c(G)$ the completion of the finitely supported functions $f \colon G \to \mathbb{C}$ for the norm

$$\|f\|_{C_c(G)} = \sup\{\|\sum_{t \in G} f(t)\pi(t)\|\}$$

where the supremum runs over all the representations $\pi \colon G \to B(H_\pi)$ of the group G which are uniformly bounded by c. We clearly have (in analogy with (6.5))

$$\|f\|_{C_c(G)} = \sup\{|\sum_{t \in G} f(t)g(t)| \mid g \in B_c(G), \, \|g\|_{B_c(G)} \leq 1\}.$$

The corresponding question for a discrete group G was answered negatively by Haagerup [H6], by showing that, while $C_c(G)$ is clearly always a Banach algebra for convolution, it is not the case for the natural predual of $M_0(G)$, when $G = F_\infty$. Hence, $M_0(G)$ is not included into the union over $c > 1$ of the spaces $B_c(G)$. However, when G is replaced by $\mathbb{N}$, the amenability of $\mathbb{Z}$ seems to be responsible for the fact that the norm $\|\ \|_{\mathcal{L}}$, which is the natural predual norm of $M_{cb}(H_1)$, is a Banach algebra norm, as mentioned above, so that the same approach does not seem to work. Hence, although it seems unlikely to be true, the question whether the inclusion $M_{cb}(H_1) \subset B_c(\mathbb{N})$ holds, remains open.

7. The similarity problem for cyclic homomorphisms on a C^*-algebra

Summary: In this chapter, we first study inequalities satisfied by any bounded linear operator u: $A \to Y$ on a C^*-algebra with values in a Banach space Y. The case when Y is another C^*-algebra is of particular interest. Then we turn to homomorphisms u: $A \to B(H)$ and prove that, if u is cyclic (= has a cyclic vector), boundedness implies complete boundedness. Hence bounded cyclic homomorphisms are similar to $*$-representations. This extends to homomorphisms with finite cyclic sets. We also include the positive solution to the similarity problem for C^*-algebras without tracial states and for nuclear C^*-algebras. Finally, we show that for a given C^*-algebra, the similarity problem and the derivation problem are equivalent.

In this chapter we will give a proof of Haagerup's solution of the similarity problem for cyclic homomorphism, cf. [H1]. Note that E. Christensen [C3] obtained positive results very close to Haagerup's. Both Christensen's and Haagerup's approaches are closely linked to an inequality for bounded operators on a C^*-algebra which was first proved in [P5].

This inequality had been conjectured earlier by J. Ringrose who had proved some special cases of it. It is related to a non-commutative version of Grothendieck's theorem (but actually it is easier to prove than the "full" non-commutative Grothendieck theorem.) We now state this inequality.

Theorem 7.1. *Let A, B be C^*-algebras and let $u: A \to B$ be a bounded operator. Then for any finite sequence x_i in A we have*

$$(7.1) \qquad \left\| \sum \frac{u(x_i)^* u(x_i) + u(x_i)u(x_i)^*}{2} \right\|^{1/2} \le 2\|u\| \left\| \sum \frac{x_i^* x_i + x_i x_i^*}{2} \right\|^{1/2}.$$

Remark. For an element x in a C^*-algebra, we define the "symmetrized modulus" $|x|$ of x as

$$(7.2) \qquad |x| = \left(\frac{x^* x + x x^*}{2} \right)^{1/2}.$$

Note that if $x = a + ib$ with a, b hermitian then

$$|x| = (a^2 + b^2)^{1/2}.$$

In the case when A and B are *both* assumed commutative, then (7.1) trivially holds (even without the factor 2). Indeed, we have

Proposition 7.2. *If A is a commutative C^*-algebra, then for any finite sequence (x_i) of elements of A*

(7.3)
$$\left\|\left(\sum_1^n |x_i|^2\right)^{1/2}\right\| = \sup_{\xi \in B_{A^*}} \left(\sum_1^n |\xi(x_i)|^2\right)^{1/2}$$

$$= \sup\left\{\left\|\sum_1^n \alpha_i x_i\right\| \ \Big| \ \alpha_i \in \mathbb{C}, \sum |\alpha_i|^2 \le 1\right\}.$$

Hence if B is another commutative C^*-algebra we have for any bounded operator $u\colon A \to B$

(7.4)
$$\left\|\left(\sum |u(x_i)|^2\right)^{1/2}\right\| \le \|u\| \left\|\left(\sum |x_i|^2\right)^{1/2}\right\|.$$

Proof: If A is a commutative C^*-algebra then there is a locally compact space Ω such that A can be identified with the space $C_0(\Omega)$ of all continuous functions on Ω which tend to zero at infinity. Then (7.3) is easy to check and (7.4) is an immediate consequence. □

Recall that a state on a C^*-algebra A is a linear functional $f\colon A \to \mathbb{C}$ such that $f \ge 0$ (i.e. $f(x) \ge 0 \ \forall \ x \ge 0$) and $\|f\| = 1$. If A has a unit it is well known that a functional f in A^* is a state iff $1 = f(1) = \|f\|_{A^*}$. Let A_h be the self-adjoint part of A. Then any state f on A defines by restriction an $\mathbb{R}$-linear form φ of norm one on A_h. Conversely given an $\mathbb{R}$-linear form φ on A_h, we may complexify φ by setting $f(a + ib) = \varphi(a) + i\varphi(b)$ for all a, b in A_h. Then (using $|\Re(f(x))| \le \|\varphi\|\|x\|$ for $x \in A$ and all its complex multiples) it is easy to check that f and φ have the same norm, so that f is a state iff $\|\varphi\| = 1$. Therefore, any positive $\mathbb{R}$-linear form of norm one on A_h uniquely extends by complexification to a state on A.

We will deduce Theorem 7.1 from the following stronger result.

Theorem 7.3. *Let $T\colon A \to H$ be an operator from a C^*-algebra A into a Hilbert space H. Then there are two states f, g on A such that*

(7.5) $$\forall \, x \in A \qquad \|T(x)\|^2 \le \|T\|^2\{f(x^*x) + g(xx^*)\}.$$

Moreover, if A has a unit there is an ultrafilter $\mathcal{U}$ on $\mathbb{N}$ and a sequence of unitary operators U_n in A and a state f on A such that for all hermitian elements a in A we have

$$\lim_{\mathcal{U}} \|T(U_n a)\|^2 \le \|T\|^2 f(a^2).$$

Proof: There are several steps. Let $\|T\| = 1$ throughout the proof.

Step 1. We first assume that A has a unit and that $\|T\| = \|T(1)\| = 1$. Let f be defined by

$$\forall \, x \in A, \qquad f(x) = \langle T(1), T(x)\rangle.$$

Then $\|f\| \le 1$ and $f(1) = 1$, so that f is a state. We claim that

$$\|Tx\|^2 \le f(x^*x + xx^*),$$

so that (7.5) holds with $g = f$. To prove this, note that

$$\forall\, a \in A_h, \quad \forall\, t \in \mathbb{R}, \qquad \|T(e^{ita})\|^2 \le 1.$$

Let $h = 1 + ita - t^2 a^2/2$. We have (when $t \to 0$)

$$\|T(h)\|^2 \le \|T(e^{ita})\|^2 + O(t^3) \le 1 + O(t^3),$$

or, equivalently, $\langle T(h), T(h) \rangle \le 1 + O(t^3)$. Developing $T(h)$, we obtain, after simplification,

$$\langle T(1), T(1) \rangle + t^2 \alpha \le 1 + O(t^3),$$

where

$$\alpha = \langle T(a), T(a) \rangle - \frac{1}{2}(\langle T(a^2), T(1) \rangle + \langle T(1), T(a^2) \rangle)$$
$$= \langle T(a), T(a) \rangle - f(a^2).$$

Letting $t \to 0$, this implies $\alpha \le 0$. Hence

$$\|T(a)\|^2 \le f(a^2).$$

Now let $x = a + ib$ be an arbitrary element of A with a, b in A_h. We have $|x|^2 = \frac{1}{2}(x^* x + x x^*) = a^2 + b^2$. Hence

$$\|T(x)\|^2 \le (\|T(a)\| + \|T(b)\|)^2 \le 2(\|T(a)\|^2 + \|T(b)\|^2)$$
$$\le 2(f(a^2) + f(b^2)) = f(x^* x + x x^*).$$

This proves the claim.

Step 2. Assume now that there is a unitary U in A such that $\|T(U)\| = \|T\| = 1$. Then the operator S defined by $S(x) = T(Ux)$ has the property in Step 1. Let f be the state associated to S as in Step 1. Define g by $g(x) = f(U^* x U)$.

A simple computation shows that

$$\|T(x)\|^2 = \|S(U^* x)\|^2 \le f(x^* x) + g(x x^*).$$

Step 3. We now treat the general case. We can always assume that A has a unit (otherwise we pass to A^{**} which always has one). Then the unit ball of A is the closed convex hull of the set of its unitary elements (cf. [Ped, p. 4]). Therefore, there are unitary operators U_n in A such that $\|T(U_n)\| \to \|T\| = 1$ when $n \to \infty$. Let $\mathcal{U}$ be a free ultrafilter on $\mathbb{N}$. We consider the ultrapowers $\hat{A} = A^{\mathbb{N}}/\mathcal{U}$, $\hat{H} = H^{\mathbb{N}}/\mathcal{U}$ and $\hat{T} = T^{\mathbb{N}}/\mathcal{U}$. Clearly $\hat{A}$ is a C^*-algebra with unit, $\hat{H}$ is a Hilbert space and $\|\hat{T}\| = \|T\| = 1$. Let $\hat{U}$ be the equivalence class in $\hat{A}$ corresponding to the sequence (U_n). Then $\hat{U}$ is unitary in $\hat{A}$ and $\|\hat{T}(\hat{U})\| = 1$. Therefore, by Step 2, there are states $\hat{f}$, $\hat{g}$ on $\hat{A}$ such that

(7.6) $$\forall\, \hat{x} \in \hat{A}, \qquad \|\hat{T}(\hat{x})\|^2 \le \hat{f}(\hat{x}^* \hat{x}) + \hat{g}(\hat{x} \hat{x}^*).$$

On the other hand, there is a canonical isometric embedding $j_A \colon A \to \hat{A}$ of A as a C^*-subalgebra of $\hat{A}$, and an isometric embedding $j_H \colon H \to \hat{H}$ such that

$j_H T = \widehat{T} j_A$. Therefore, letting f, g be the restrictions of $\hat{f}, \hat{g}$ to A (i.e., $f = j_A^* \hat{f}$, $g = j_A^* \hat{g}$) we deduce (7.5) from (7.6).

Moreover, if A has a unit by the first step of the proof we can assume that for any hermitian $\hat{a}$ in $\hat{A}$ we have

$$\|\widehat{T}(\widehat{U} \hat{a})\|^2 \le \hat{f}(\hat{a}^2).$$

Taking $\hat{a} = j_A(a)$ with a hermitian in A we obtain

$$\|\widehat{T}(\widehat{U} \hat{a})\|^2 = \lim_{\mathcal{U}} \|T(U_n a)\|^2 \le \hat{f}(\hat{a}^2) = \hat{f}(\hat{a}^2) = f(a^2)$$

which completes the proof.   □

Note that if (7.5) holds and if $\|T\| \le 1$, then there is a state $\varphi (\varphi = (1/2)(f + g))$ such that

$$\forall\, x \in A \qquad \|T(x)\|^2 \le 2\varphi(x^* x + x x^*) = 4\varphi(|x|^2)$$

with the modulus as defined in (7.2). We note in passing that the existence of such a φ is equivalent to the validity of a certain inequality, as follows.

Proposition 7.4. *Let* $T \colon A \to Y$ *be an operator with values in a Banach space* Y, *and let* C *be some constant. The following assertions are equivalent:*

(i) *There is a state* f *on* A *such that*

$$\forall\, x \in A, \qquad \|Tx\| \le C(f(|x|^2))^{1/2}.$$

(ii) *For all finite sequences* (x_i) *in* A, *we have*

$$\sum \|T x_i\|^2 \le C^2 \left\| \sum |x_i|^2 \right\|.$$

Proof: (i) $\Rightarrow$ (ii) is obvious. Conversely assume (i). We will use the Hahn-Banach theorem as stated above in Corollary 3.2. Let $A_+ = \{x \in A \mid x \ge 0\}$ let $p(x) = \|x\|$ on A and for all $x \ge 0$ let

$$q(x) = \sup \left\{ \sum \|T x_i\|^2 \right\}$$

where the supremum runs over all finite sequences (x_i) such that $C^2 \sum |x_i|^2 \le x$. By our assumption (ii) we have $q(x) \le p(x)$ and q clearly is superlinear. Hence by Corollary 3.2 there is a IR-linear form f on A such that $f(x) \le \|x\|$ for all x in A and $q(x) \le f(x)$ for all $x \ge 0$. Obviously $f \ge 0$ hence after normalization and complexification (see the remarks before Theorem 7.3) f is a state and it clearly satisfies (i).   □

Proof of Theorem 7.1. We may assume that B is a C^*-subalgebra of $B(H)$ for some Hilbert space H and $\|u\| = 1$. Then choose ξ in H with $\|\xi\| = 1$, and apply Theorem 7.3 to the operator $T \colon A \to H$ defined by $T(x) = u(x)\xi$ for x in A. We then find a state ϕ on A such that

$$\sum \|u(x_i)\xi\|^2 \le 4\phi \left(\sum |x_i|^2 \right) \le 4 \left\| \sum |x_i|^2 \right\|.$$

By a similar reasoning, we find, for all y_i in A,

$$(7.7) \qquad \sum \|u(y_i^*)^*\xi\|^2 \le 4 \left\| \sum |y_i|^2 \right\|.$$

Hence, taking $y_i = x_i^*$ in (7.7) and adding we get

$$(7.8) \qquad \frac{1}{2} \left\langle \sum u(x_i)^* u(x_i)\xi + \sum u(x_i)u(x_i)^*\xi, \xi \right\rangle \le 4 \left\| \sum |x_i|^2 \right\|.$$

Finally, taking the supremum in (7.8) over all possible ξ, we obtain (7.1).   □

We now turn to Haagerup's theorem.

Theorem 7.5. *Let A be a C^*-algebra, H a Hilbert space and let $u\colon A \to B(H)$ be a bounded homomorphism. Assume that u has a cyclic vector ξ, i.e. a vector ξ in H such that*

$$\overline{u(A)\xi} = H.$$

Then u is c.b. and

$$\|u\|_{cb} \le \|u\|^4.$$

Therefore (by Corollary 4.4), in the unital case there is an isomorphism

$$S\colon H \to H \quad \text{with} \quad \|S\|\,\|S^{-1}\| \le \|u\|^4$$

such that $a \to S^{-1}u(a)S$ is a (C^*-algebra) representation.

We will use a series of simple lemmas.

Lemma 7.6. *If $u\colon A \to B(H)$ is a homomorphism then there is a homomorphism $\tilde{u}\colon A^{**} \to B(H)$ extending u with $\|\tilde{u}\| = \|u\|$, and moreover continuous for the topologies $\sigma(A^{**}, A^*)$ and $\sigma(B(H), B(H)_*)$. (Recall, as is well known, that A^{**} is a unital von Neumann algebra.)*

Proof: Let $v\colon B(H)_* \to A^*$ be the restriction of u^* to $B_*(H)$ and let $\tilde{u} = v^*$. It is not too hard to check that $\tilde{u}$ is a homomorphism, and it obviously has the announced properties.   □

The next lemma shows that when u satisfies additional algebraic identities Theorem 7.1 can be significantly improved by rather simple manipulations.

Lemma 7.7. *If $u\colon A \to B(H)$ is a bounded homomorphism then for all finite sequences (x_i) in A we have*

$$(7.9) \qquad \left\| \sum u(x_i)^* u(x_i) \right\| \le \|u\|^4 \left\| \sum x_i^* x_i \right\|.$$

Proof: By the preceding lemma, we can assume that A is a von Neumann algebra and in particular that A has a unit. Let $\xi \in H$ be arbitrary with $\|\xi\| = 1$. Let $T\colon A \to H$ be the operator defined by $T(a) = u(a)\xi$. Then there are $\mathcal{U}$, (U_n) and f as in the second part of Theorem 7.3 such that for any hermitian a in A we have

$$(7.10) \qquad \lim_{\mathcal{U}} \|u(U_n a)\xi\|^2 \le \|u\|^2 f(a^2).$$

Observe that for all a, b in A we have

(7.11) $\qquad u(ab)^* u(ab) = u(b)^* u(a)^* u(a) u(b) \le \|u(a)\|^2 u(b)^* u(b).$

Now let x be arbitrary in A. Since A is a von Neumann algebra, there is a polar decomposition $x = vy$ with $v, y \in A$,

$$y = (x^* x)^{1/2} \quad \text{and} \quad \|v\| \le 1.$$

We have $x = (vU_n^*)U_n y$ hence by (7.11) since $\|u(vU_n^*)\| \le \|u\|$ we have

$$u(x)^* u(x) \le \|u\|^2 u(U_n y)^* u(U_n y)$$

hence

$$\|u(x)\xi\|^2 \le \|u\|^2 \|u(U_n y)\xi\|^2$$

which implies by (7.10)

$$\forall\, x \in A \qquad \|u(x)\xi\|^2 \le \|u\|^4 f(y^2) = \|u\|^4 f(x^* x).$$

Therefore for any finite sequence $x_1, \ldots, x_n$ in A

$$\left\langle \sum u(x_i)^* u(x_i)\xi, \xi \right\rangle = \sum \|u(x_i)\xi\|^2 \le \|u\|^4 f\left(\sum x_i^* x_i\right) \le \|u\|^4 \left\|\sum x_i^* x_i\right\|$$

and taking the sup over all ξ with $\|\xi\| \le 1$ we obtain (7.9).

The next lemma is quite elementary.

Lemma 7.8. *Let* $(a_i)_{i \le n}, (b_j)_{j \le n}$ *be elements in a* C^*-*algebra* A. *Then*

(7.12) $\qquad \|(a_i b_j^*)\|_{M_n(A)} \le \left\|\left(\sum a_i^* a_i\right)^{1/2}\right\| \left\|\left(\sum b_j^* b_j\right)^{1/2}\right\|.$

Proof: Let $\alpha = (a_i b_j^*) \in M_n(A)$. We have $\alpha = \beta\gamma$ with

$$\beta = \begin{pmatrix} a_1 \\ \vdots & \text{\Large 0} \\ a_n \end{pmatrix} \quad \text{and} \quad \gamma = \begin{pmatrix} b_1^* & \cdots & b_n^* \\ & \text{\Large 0} \end{pmatrix}$$

hence $\|\alpha\| \le \|\beta\|\,\|\gamma\|$ and this yields (7.12). $\qquad\qquad\square$

Combining Lemmas 7.8 and 7.7 we find

Lemma 7.9. *If* $u: A \to B(H)$ *is a bounded homomorphism then for all* $a_1, \ldots,$ $a_n, b_1, \ldots, b_n$ *in* A *we have*

$$\|(u(a_i b_j^*))\|_{M_n(B(H))} \le \|u\|^4 \left\|\sum a_i^* a_i\right\|^{1/2} \left\|\sum b_j^* b_j\right\|^{1/2}.$$

Proof: We simply write $u(a_i b_j^*) = u(a_i) u_*(b_j)^*$ where u_* is the homomorphism defined by

$$\forall\, x \in A \qquad u_*(x) = u(x^*)^*.$$

Note that $\|u_*\| = \|u\|$. Then the result follows by Lemma 7.8 from Lemma 7.7 applied twice (to u and to u_*). $\qquad\qquad\square$

Finally we come to the key step.

Lemma 7.10. *If* $u \colon A \to B(H)$ *is a bounded homomorphism and if* (x_i) *and* (y_j) *are finite sequences in* A *such that*

$$\sum y_i^* y_i \leq \sum x_j^* x_j$$

then we have

(7.13)
$$\sum u(y_i)^* u(y_i) \leq \|u\|^8 \sum u(x_j)^* u(x_j).$$

Proof: Without loss of generality we may assume that A has a unit and that $a = \left(\sum x_j^* x_j \right)^{1/2}$ is invertible (just add εI to the family (x_j) for some $\varepsilon > 0$ and let ε tend to zero in the end). Moreover, by completing (y_i) if necessary we can even assume $\sum y_i^* y_i = \sum x_j^* x_j = a^2$. Then let $b_j = x_j a^{-1}$ and $a_i = y_i a^{-1}$. Then $\sum b_j^* b_j = a^{-1} a^2 a^{-1} = I$ and similarly $\sum a_i^* a_i = I$. Hence by Lemma 7.9 if we define $a_{ij} = a_i b_j^*$ we have

(7.14)
$$\|(u(a_{ij}))\|_{M_n(B(H))} \leq \|u\|^4.$$

On the other hand it is easy to check that $y_i = \sum_j a_{ij} x_j$ hence we have $u(y_i) = \sum_j u(a_{ij}) u(x_j)$ hence by (7.14) for any ξ in H

$$\sum \|u(y_i)\xi\|^2 \leq \|u\|^8 \sum_j \|u(x_j)\xi\|^2$$

which is equivalent to (7.13).   □

Proof of Theorem 7.5. Let (a_{ij}) be arbitrary in the unit ball of $M_n(A)$. To show $\|u\|_{cb} \leq \|u\|^4$ it suffices to show that for all $h_1, \ldots, h_n$ in H we have

(7.15)
$$\sum_i \left\| \sum_j u(a_{ij}) h_j \right\|^2 \leq \|u\|^8 \sum \|h_j\|^2.$$

Since ξ is cyclic it suffices to show (7.15) for (h_j) of the form $h_j = u(x_j)\xi$ with $x_1, \ldots, x_n$ arbitrary in A. But then

$$\sum_j u(a_{ij}) u(x_j)\xi = u(y_i)\xi \quad \text{with} \quad y_i = \sum_j a_{ij} x_j$$

and since $\|(a_{ij})\|_{M_n(A)} \leq 1$ we clearly have $\sum y_i^* y_i \leq \sum x_j^* x_j$. Hence by Lemma 7.10

$$\sum \|u(y_i)\xi\|^2 = \left\langle \sum u(y_i)^* u(y_i)\xi, \xi \right\rangle$$
$$\leq \|u\|^8 \left\langle \sum u(x_j)^* u(x_j)\xi, \xi \right\rangle = \|u\|^8 \sum \|u(x_j)\xi\|^2,$$

which yields (7.15).   □

A simple variant of Theorem 7.5 gives

Theorem 7.11. *If a bounded homomorphism $u\colon A \to B(H)$ on a C^*-algebra has a finite cyclic set, i.e. if there is a subset $\{\xi_1, \ldots, \xi_n\} \subset H$ such that*

(7.16)
$$\overline{\mathrm{span}}\left(\bigcup_{k=1}^{n} u(A)\xi_i\right) = H,$$

then u is c.b. and we have

$$\|u\|_{cb} \le \|I_{M_n} \otimes u\|^4_{M_n(A)\to M_n(B(H))}.$$

Proof: Let us denote $u_n = I_{M_n} \otimes u\colon M_n(A) \to M_n(B(H))$. Clearly, u_n is a bounded homomorphism on $M_n(A)$ and $\xi = (\xi_1, \ldots, \xi_n)$ is a cyclic vector for u_n. Hence by Theorem 7.5 we have $\|u_n\|_{cb} \le \|u_n\|^4$. But we have

$$u = q u_n j$$

where $j\colon A \to M_n(A)$ is the $*$-representation $a \to e_{11} \otimes a$ and where the map $q\colon M_n(B(H)) \to B(H)$ is defined by $q((a_{ij})) = a_{11}$. Clearly $\|q\|_{cb} \le 1$ and $\|j\|_{cb} \le 1$, hence $\|u\|_{cb} \le \|u_n\|_{cb} \le \|u_n\|^4$. $\qquad\square$

Returning to the similarity problem in the group setting (Revised Problem 0.1), we note the following consequence of Theorem 7.11.

Corollary 7.12. *Let $\pi\colon G \to B(H)$ be a uniformly bounded representations on a discrete group G. Assume that π has a finite cyclic set. Then the following are equivalent*

(i) *Every coefficient of π is in $B(G)$.*
(ii) *π is similar to a unitary representation.*

Proof: By a coefficient of π, we mean of course a function $f\colon G \to \mathbb{C}$ of the form $f_{\xi,\eta}(t) = \langle \pi(t)\xi, \eta \rangle$ with $\xi, \eta \in H$. Assume (i). We claim that π extends in the obvious way to a bounded unital homomorphism $u_\pi\colon C^*(G) \to B(H)$. Indeed, the mapping $(\xi, \eta) \to f_{\xi,\eta}$ must clearly be a bounded bilinear form from $H \times H$ into $B(G)$, hence for some constant C we have

$$\forall \xi, \eta \in H \qquad\qquad \|f_{\xi,\eta}\|_{B(G)} \le C\|\xi\|\,\|\eta\|.$$

This implies that for any x in $\ell_1(G)$ we have

$$\left\|\sum x(t)\pi(t)\right\|_{B(H)} = \sup\left|\sum_{t\in G} x(t)f_{\xi,\eta}(t)\right|$$
$$\le C\|x\|_{B(G)^*} = \|x\|_{C^*(G)}.$$

Therefore $x \to \sum x(t)\pi(t)$ extends as announced to a unital homomorphism u_π on $C^*(G)$ with $\|u_\pi\| \le C$. If π has a finite cyclic set so does u_π, hence by Theorem 7.11, u_π is similar to a $*$-representation, therefore (equivalently in fact) π is similar to a unitary representation. This shows (i) $\Rightarrow$ (ii). The converse is obvious. $\qquad\square$

Recall that a homomorphism $u: A \to B(H)$ is called nondegenerate if $\bigcup_{\xi \in H} u(A)\xi$ is dense in H. If the algebra A is "big enough" we can use a trick to replace finite cyclic sets by cyclic vectors. Indeed, assume that there are isometries $s_1, \ldots, s_n$ in A such that

(7.17) $$s_i^* s_j = \delta_{ij} I \qquad 1 \leq i, j \leq n.$$

Then, if $\{\xi_1, \ldots, \xi_n\}$ is cyclic for u (i.e. (7.16) holds) then the vector

(7.18) $$\xi = \sum_1^n u(s_i)\xi_i$$

is cyclic for u since we have

$$\sum_1^n u(A)\xi_i = u(A)\xi.$$

Indeed note that for all $a_1, \ldots, a_n$ in A we have $\sum u(a_i)\xi_i = u(\sum a_i s_i^*)\xi$. This yields the following result.

Proposition 7.13. *Consider a C^*-algebra A. Assume that for every $n \geq 1$, A contains isometries $s_1, \ldots s_n$ satisfying (7.17). Then every nondegenerate bounded homomorphism $u: A \to B(H)$ is c.b. and we have $\|u\|_{cb} \leq \|u\|^4$.*

Proof: Choose a finite subset $J = \{\xi_1, \ldots, \xi_n\}$ arbitrarily in H. Let $H_J \subset H$ be the closure of $\sum_1^n u(A)\xi_i$. Let $u_J: A \to B(H_J)$ be defined by $u_J(a) = u(a)_{|H_J}$. Then u_J is a bounded homomorphism admitting $\{\xi_1, \ldots, \xi_n\}$ as cyclic set. By the preceding remark (using (7.18)) it actually admits a cyclic vector hence by Theorem 7.5 we have $\|u_J\|_{cb} \leq \|u\|^4$. Then, since the family $\{H_J \mid J \subset H, J$ finite$\}$ of subspaces of H is directed by inclusion with union dense in H (recall u is assumed nondegenerate), we have

$$\|u\|_{cb} = \sup_J \|u_J\|_{cb}$$

hence by the first part of the proof $\|u\|_{cb} \leq \|u\|^4$.   □

Corollary 7.14. *Let A be a C^*-algebra without tracial states. Then every nondegenerate bounded homomorphism $u: A \to B(H)$ is similar to a $*$-representation (or equivalently is completely bounded.).*

Proof: Using Lemma 7.6, we can replace A be its bidual A^{**} which (by assumption) has no normal tracial states. It follows that A^{**} is a property infinite von Neumann algebra. In particular (by the "halving" lemma, see e.g. [KR, p. 412]) A^{**} satisfies the assumption of the preceding proposition, whence the conclusion.   □

Remark. Note that the preceding result applies in the case $A = B(\mathcal{H})$.

On the other hand, a typical example of a separable C^*-algebra satisfying the assumption of Proposition 7.13 is the Cuntz algebra ([Cu]). More generally Corollary 7.14 applies to all simple, infinite C^*-algebras in the sense of [Cu]. However, since the Cuntz algebra is nuclear, this also can be deduced from the next theorem.

We need to recall an equivalent definition of the now classical notion of nuclearity for a C^*-algebra.

Definition 7.15. *A C^*-algebra A is called nuclear if there exists a net of finite rank maps V_α: $A \to A$ converging pointwise to the identity such that each map V_α admits a factorization of the form*

$$V_\alpha: A \xrightarrow{S_\alpha} M_{n_\alpha} \xrightarrow{T_\alpha} A$$

with T_α completely positive, S_α completely bounded and

(7.19) $$\|S_\alpha\|_{cb} \, \|T_\alpha\| \leq 1.$$

For such algebras, the similarity problem is easily solved as follows.

Theorem 7.16. *Every bounded homomorphism u: $A \to B(H)$ defined on a nuclear C^*-algebra is c.b. with $\|u\|_{cb} \leq \|u\|^2$ (and therefore by Corollary 4.4 is similar to a $*$-representation in the unital case).*

For the proof of Theorem 7.16, we will use the following statement which is the completely positive counterpart of Corollary 3.11.

Lemma 7.17. *Let $n \geq 1$. Let A be any C^*-algebra.*

(i) *A mapping T: $M_n \to A$ is completely positive iff there is an element (a_{ij}) in $M_n(A)$ such that*

$$T(e_{ij}) = \sum_k a_{ik}^* a_{jk} \qquad \forall \, i, j \leq n.$$

(ii) *Consider (a_{ij}) and (b_{ij}) in $M_n(A)$. Let W: $M_n \to A$ be the linear mapping defined by*

$$W(e_{ij}) = \sum_k a_{ik}^* b_{jk}.$$

Then

$$\|W\|_{cb} \leq \left\| \sum_{ij} a_{ij}^* a_{ij} \right\|^{1/2} \left\| \sum_{ij} b_{ij}^* b_{ij} \right\|^{1/2}.$$

Proof: (i) First we observe that the matrix $e = (e_{ij})$ (formed of the so-called matrix units) is positive in $M_n(A)$. Indeed we have

$$(ee^*)_{ij} = \sum_k e_{ik} e_{jk}^* = \sum_k e_{ik} e_{kj} = n \, e_{ij}$$

hence $e = (1/n)ee^* \geq 0$. Therefore, if T: $M_n \to A$ is completely positive, the matrix $(T(e_{ij}))$ must be positive in $M_n(A)$, hence it can be written as a^*a for some a in $M_n(A)$ and this yields (i).

To verify (ii), assume $A \subset B(H)$ and let $\hat{H} = \ell_2^n(H)$ and let y_i: $H \to \hat{H}$ and x_j: $H \to \hat{H}$ be the operators defined by

$$\forall \, h \in H \qquad y_i(h) = (a_{i1}h, \ldots, a_{in}h)$$
$$x_j(h) = (b_{j1}h, \ldots, b_{jn}h).$$

Then, we have $W(e_{ij}) = \sum_k a_{ik}^* b_{jk} = y_i^* x_j$ and moreover

$$\left\| \sum y_i^* y_i \right\| = \left\| \sum_{ij} a_{ij}^* a_{ij} \right\|, \qquad \left\| \sum x_j^* x_j \right\| = \left\| \sum_{ij} b_{ij}^* b_{ij} \right\|.$$

Hence part (ii) follows from Corollary 3.11.   □

We note the following consequence of Lemma 7.17.

Proposition 7.18. *Let A be an arbitrary C^*-algebra and let $u\colon A \to B(H)$ be a bounded homomorphism. Then, for any $n \geq 1$, for any completely positive mapping $T\colon M_n \to A$ we have*

$$\|uT\|_{cb} \leq \|u\|^4 \|T\|.$$

Proof: Let (a_{ij}) be as in the first part of Lemma 7.17. Then we have (recall $u_*(x) = u(x^*)^*$)

$$uT(e_{ij}) = \sum_k u(a_{ik}^*) u(a_{jk})$$

$$= \sum_k u_*(a_{ik})^* u(a_{jk})$$

hence by part (ii) in Lemma 7.17

$$\|uT\|_{cb} \leq \left\| \sum_{ij} u_*(a_{ij})^* u_*(a_{ij}) \right\|^{1/2} \left\| \sum_{ij} u(a_{ij})^* u(a_{ij}) \right\|^{1/2}$$

hence by Lemma 7.7 applied both to u and u_*

$$\|uT\|_{cb} \leq \|u\|^4 \left\| \sum_{ij} a_{ij}^* a_{ij} \right\| = \|u\|^4 \left\| T\left(\sum_i e_{ii} \right) \right\| \leq \|u\|^4 \|T\|.$$

□

First proof of Theorem 7.16. This proof will give a slightly worse bound on $\|u\|_{cb}$ than announced. Let S_α, T_α be as in Definition 7.15. Since $T_\alpha S_\alpha \to Id_A$ pointwise, we clearly have

$$\|u\|_{cb} \leq \sup_\alpha \|uT_\alpha S_\alpha\|_{cb}$$

hence

$$\leq \sup_\alpha \|uT_\alpha\|_{cb} \|S_\alpha\|_{cb}$$

hence by the preceding proposition and by (7.19)

$$\leq \sup_\alpha \{ \|u\|^4 \|T_\alpha\| \ \|S_\alpha\|_{cb} \} \leq \|u\|^4.$$

For the more precise bound $\|u\|_{cb} \leq \|u\|^2$, see the second proof below.   □

We start by a very particular case of Theorem 7.16.

Lemma 7.19. *Let $u: M_n \to B(H)$ be a unital homomorphism. Then*

$$(7.20) \qquad \|u\|_{cb} \le \|u\|^2.$$

More generally, this remains valid if we replace M_n by any finite dimensional von Neumann algebra.

Proof: Let G be the group of unitary matrices, considered as a topological subspace of M_n. Let π be the restriction of u to G. Then π is a uniformly bounded group representation with $\sup_{t \in G} \|\pi(t)\| = \|u\|$. Note that G is compact hence amenable, and π is continuous. Therefore, by Theorem 0.6, there is a similarity S on H with $\|S\|\|S^{-1}\| \le \|u\|^2$, such that $S^{-1}\pi(.)S$ is a unitary representation. Clearly, this means equivalently that $S^{-1}u(.)S$ is a $*$-representation, hence a complete contraction. It follows immediately, by the (trivial) converse direction in Theorem 3.6, that $\|u\|_{cb} \le \|S\|\|S^{-1}\| \le \|u\|^2$.

The same argument applies when M_n is replaced by a finite dimensional von Neumann algebra.   □

Remark. A slightly weaker estimate than (7.20), namely

$$\|u\|_{cb} \le \|u\|^4,$$

can be deduced from Lemma 7.7. Indeed, let $y_i^* = u(e_{i1})$ and $x_j = u(e_{1j})$. Then $u(e_{ij}) = u(e_{i1})u(e_{1j}) = y_i^* x_j$. Therefore, by Corollary 3.11 we have

$$\|u\|_{cb} \le \left\| \sum y_i^* y_i \right\|^{1/2} \left\| \sum x_j^* x_j \right\|^{1/2}$$

hence by Lemma 7.7 applied to u and $u_*: x \to u(x^*)^*$ (and observing $\sum e_{i1}e_{i1}^* = \sum e_{1j}^* e_{1j} = I$) we find

$$\|u\|_{cb} \le \|u\|^4 \left\| \sum e_{i1}e_{i1}^* \right\|^{1/2} \left\| \sum e_{1j}^* e_{1j} \right\|^{1/2} \le \|u\|^4.$$

Second proof of Theorem 7.16. A very simple proof can be given using Connes's celebrated result ([Co]) that the bidual A^{**} is hyperfinite, i.e. we have a directed net $N_\alpha \subset A^{**}$ of finite dimensional von Neumann algebras such that their union $\cup N_\alpha$ is dense in A^{**} for the $\sigma(A^{**}, A^*)$-topology. Moreover, for each α there is a projection $P_\alpha: A^{**} \to N_\alpha$ with $\|P_\alpha\|_{cb} \le 1$, and P_α tends to the identity pointwise in the $\sigma(A^{**}, A^*)$-topology.

Note that the case when A is finite dimensional is settled by Lemma 7.19.

Now for the general case we use Connes's result. (Note in passing that we can assume without loss of generality that A is separable.) Consider the homomorphism $\tilde{u}: A^{**} \to B(H)$ as in Lemma 7.6. By Lemma 7.19, we know that $\|\tilde{u}_{|N_\alpha}\|_{cb} \le \|u\|^2$. Clearly we have $\tilde{u}P_\alpha \to \tilde{u}$ pointwise with respect to the $\sigma(B(H), B(H)_*)$ topology on $B(H)$. It is easy to check that this implies $\|\tilde{u}\|_{cb} \le \sup \|\tilde{u}P_\alpha\|_{cb}$ and since $\tilde{u}P_\alpha = \tilde{u}_{|M_\alpha}P_\alpha$ we have $\|\tilde{u}P_\alpha\|_{cb} \le \|\tilde{u}_{|M_\alpha}\|_{cb} \le \|u\|^2$. Hence we conclude that $\|\tilde{u}\|_{cb} \le \|u\|^2$, and a fortiori since $u = \tilde{u}_{|A}$ we have $\|u\|_{cb} \le \|u\|^2$.   □

We note the following interesting application of Theorem 7.11.

Theorem 7.20. *The following properties of a C^*-algebra A are equivalent.*

(i) *For any H, every nondegenerate bounded homomorphism $u\colon A \to B(H)$ is similar to a $*$-representation.*

(ii) *There is a nondecreasing function $f\colon \mathbb{R}_+ \to \mathbb{R}_+$ such that, for any H, any nondegenerate bounded homomorphism $u\colon A \to B(H)$ satisfies $\|u\|_{cb} \le f(\|u\|)$.*

(iii) *There is a nondecreasing function $f\colon \mathbb{R}_+ \to \mathbb{R}_+$ with the following property: for any Hilbert space $\widehat{H}$, for any $*$-representation $\pi\colon A \to B(\widehat{H})$ and for any invertible operator $S\colon \widehat{H} \to \widehat{H}$, the homomorphism $\pi_S\colon A \to B(\widehat{H})$ defined by $\pi_S(x) = S^{-1}\pi(x)S$ satisfies $\|\pi_S\|_{cb} \le f(\|\pi_S\|)$.*

Note: Recall that it is an open problem (cf. Problem 0.2) whether the preceding (equivalent) properties are true for *any* C^*-algebra A.

Proof: The implication (i) $\Rightarrow$ (ii) is routine (left to the reader) and (ii) $\Rightarrow$ (iii) is trivial. We will show (iii) $\Rightarrow$ (i). Assume (iii). Let u be as in (i). Let $(\xi_j)_{j\in J}$ be any finite subset of H. We consider the invariant subspace H_J generated by $(\xi_j)_{j\in J}$, namely we set

$$H_J = \overline{span}\{u(A)\xi_j \mid j \in J\}.$$

Let $u_J\colon A \to B(H_J)$ be the "restriction" of u to H_J (i.e. $u_J(a) = u(a)_{|H_J} \in B(H_J)$). Let $n = card(J)$. Obviously, u_J admits a cyclic set with n-elements. Hence by Haagerup's theorem (i.e. by Theorem 7.11) we know that u_J is c.b., so that by Theorem 7.11, there is $S_J\colon H_J \to H_J$ invertible such that $S_J^{-1}u_J(\cdot)S_J$ is a $*$-representation. In other words, u_J is of the special form considered in (iii), and since we assume (iii) we have

$$\|u_J\|_{cb} \le f(\|u_J\|) \le f(\|u\|).$$

But now a moment of thought shows that

$$\|I_{M_n} \otimes u\|_{M_n(A)\to M_n(B(H))} \le \sup_{|J|=n} \|u_J\|_{cb}$$

hence

$$\|u\|_{cb} \le \sup\{\|u_J\|_{cb} \mid |J| < \infty\} \le f(\|u\|).$$

This shows that u is c.b. and completes the proof that (iii) $\Rightarrow$ (i). $\square$

The next result was observed by Kirchberg [Ki]. It shows that for a given C^*-algebra the similarity problem (Problem 0.2) and the derivation problem (Problem 0.2') are essentially equivalent. We need to enlarge a bit the framework: let $\pi\colon A \to B(\widehat{H})$ be a $*$-representation on a C^*-algebra A, we will say that a linear mapping $\delta\colon A \to B(\widehat{H})$ is a derivation relative to π if we have $\delta(ab) = \pi(a)\delta(b) + \delta(a)\pi(b)$ for all a,b in A. For example, given any T in $B(\widehat{H})$ the mapping δ_T defined by

$$\delta_T(a) = [T, \pi(a)] = T\pi(a) - \pi(a)T$$

clearly is a derivation relative to π. We again call these derivations (relative to π) inner.

Theorem 7.21. *Let A be a C^*-algebra. The properties in Theorem 7.20 are equivalent to the following additional ones.*

(iv) *For any H and for any $*$-representation $\pi\colon A \to B(H)$, any bounded derivation $\delta\colon A \to B(H)$ relative to π is c.b.*

(v) *There is a constant K such that, for any $\widehat{H}$, for any $*$-representation $\pi\colon A \to B(H)$, any inner derivation $\delta\colon A \to B(H)$ relative to π satisfies $\|\delta\|_{cb} \leq K\|\delta\|$.*

Proof: We first check (ii) $\Rightarrow$ (iv). As already mentioned above, if δ is a derivation relative to π then the mapping u defined by

$$u(a) = \begin{pmatrix} \pi(a) & \delta(a) \\ 0 & \pi(a) \end{pmatrix}$$

is a homomorphism from A to $B(H \oplus H)$. Moreover we have $\|u\| \leq 1 + \|\delta\|$ and $\|\delta\|_{cb} \leq \|u\|_{cb}$. Hence if (ii) holds we find $\|\delta\|_{cb} \leq f(\|u\|) \leq f(1 + \|\delta\|)$, which yields (iv).

The implication (iv) $\Rightarrow$ (v) is easy. Indeed, if (v) fails we have a sequence of derivations δ_n relative respectively to $*$-representations π_n such that $\|\delta_n\|_{cb} \geq n\|\delta_n\|$. Taking

$$\delta = \bigoplus_n \delta_n \quad \text{and} \quad \pi = \bigoplus_n \pi_n$$

we obtain easily a contradiction to (iv). This shows that (iv) $\Rightarrow$ (v).

Now assume (v). We will show that (iii) holds. Consider a $*$-representation $\pi\colon A \to B(H)$ and a similarity $S\colon H \to H$ as in (iii). By Theorem 4.3 and the remark after Corollary 4.4, we can assume without loss of generality that S satisfies

$$(7.21) \qquad\qquad \|S\|\,\|S^{-1}\| = \|\pi_S\|_{cb}.$$

(Indeed, by Theorem 4.3 there is S_1 with $\|S_1\|\,\|S_1^{-1}\| = \|\pi_S\|_{cb}$ such that $a \to S_1 S^{-1} \pi(a) S S_1^{-1}$ is $*$-representation, which implies that $SS_1^{-1} = CU$ for some unitary U and for some C in $\pi(A)'$. Thus if we replace S by US_1 we find $\pi_S = \pi_{US_1}$ and $\|US_1\|\,\|(US_1)^{-1}\| = \|\pi_S\|_{cb}$.) Moreover (by the polar decomposition of S) we can always assume that S is Hermitian positive (*i.e.* $S \geq \varepsilon I$ for some $\varepsilon > 0$).
We have

$$(7.22) \qquad \forall a \in A \qquad\qquad \|S^{-1}\pi(a)S\| \leq \|\pi_S\|\,\|a\|,$$

hence taking adjoints

$$(7.23) \qquad \|S\pi(a)S^{-1}\| = \|(S^{-1}\pi(a^*)S)^*\| \leq \|\pi_S\|\,\|a^*\| = \|\pi_S\|\,\|a\|,$$

Fix a in A. Now consider the entire analytic function

$$f(z) = S^z \pi(a) S^{-z}$$

with values in $B(H)$.

By the maximum principle we have by (7.22) and (7.23)

$$\|f(z)\| \le \|\pi_S\| \|a\|$$

for all z in the set $\{-1 \le Re(z) \le 1\}$, and a fortiori for all z in the closed unit disc. Hence by Cauchy's formula we have

$$\|f'(0)\| \le \|\pi_S\| \|a\|$$

or equivalently

$$\|\text{Log}(S)\, \pi(a) - \pi(a)\, \text{Log}(S)\| \le \|\pi_S\| \|a\|.$$

Let $\delta(a) = \text{Log}\,(S)\, \pi(a) - \pi(a)\, \text{Log}(S)$. Clearly δ is an inner derivation relative to π satisfying (by the last estimate)

$$(7.24) \qquad\qquad \|\delta\| \le \|\pi_S\|.$$

Now, let $a \in A$ be fixed again. Observe that $f'(z) = S^z \delta(a) S^{-z}$ for all z in $\mathbb{C}$. In particular, we have

$$S\pi(a)S^{-1} = f(1) = f(0) + \int_0^1 f'(t)dt.$$

Hence

$$\pi_S(a) = \pi(a) + \int_0^1 S^t \delta(a) S^{-t} dt,$$

which implies obviously

$$\|\pi_S\|_{cb} \le \|\pi\|_{cb} + \int_0^1 \|S^t\| \|\delta\|_{cb} \|S^{-t}\| dt$$

$$\le 1 + \|\delta\|_{cb} \int_0^1 (\|S\| \|S^{-1}\|)^t dt$$

$$= 1 + \|\delta\|_{cb} \left(\frac{\|S\| \|S^{-1}\| - 1}{\text{Log}(\|S\| \cdot \|S^{-1}\|)} \right).$$

But now since we have adjusted S so that $\|S\| \|S^{-1}\| = \|\pi_S\|_{cb}$, this last estimate yields:

$$\|\pi_S\|_{cb} \le 1 + \|\delta\|_{cb} \left(\frac{\|\pi_S\|_{cb} - 1}{\text{Log}(\|\pi_S\|_{cb})} \right)$$

hence (since we can assume $\|\pi_S\|_{cb} > 1$)

$$\text{Log}\|\pi_S\|_{cb} \le \|\delta\|_{cb}$$

and finally by (7.21) since we assume (v)

$$\|\pi_S\|_{cb} \le \exp(K\|\pi_S\|).$$

This shows that (v) implies (iii) with $f(t) = \exp(Kt)$.   □

Remark. Note that the preceding proof of (v) $\Rightarrow$ (iii) actually remains valid for any fixed $*$-representation $\pi\colon A \to B(H)$.

Notes and Remarks on Chapter 7

General references on C^*-algebras are [Ped] and [KR]. (Warning: all C^*-algebras are *assumed* unital in [KR].)

It might be worthwhile to start by clarifying the respective rôle of the assumptions "unital" and "nondegenerate".

For simplicity, we have restricted our discussion of similarity to the unital case, but as is well known, it is easy to reduce the general case to the unital one, as follows:

First observe that a nondegenerate homomorphism $u\colon B \to B(H)$ on a unital C^*-algebra B must necessary take the unit of B to the identity of H.

On the other hand, by Lemma 7.6, any bounded homomorphism $u\colon A \to B(H)$ on a C^*-algebra A extends to one $\tilde{u}\colon A^{**} \to B(H)$ on A^{**} which is a unital algebra. If u is assumed nondegenerate, so is its extension $\tilde{u}$, therefore by the preceding observation, $\tilde{u}$ is necessarily unit preserving.

Finally, applying Lemma 7.6 to $M_n(A)$ we find that $\|\tilde{u}\|_{cb} = \|u\|_{cb}$. Thus, (applying Corollary 4.4 to $\tilde{u}$) we conclude that a bounded nondegenerate homomorphism $u\colon A \to B(H)$ on a (not necessarily unital) C^*-algebra A is cb iff it is similar to a $*$-representation.

This allows, for instance, to remove the assumption that u is unital in the conclusions of Theorems 7.5 and 7.16.

Except for the constant, Theorem 7.1 was first proved in [P5]. A new proof of it was given by Haagerup [H1, Appendix] with a better constant as stated above. A different proof also appears in [LuPP]. Generalizations of Theorem 7.1 can be found in [HP1]. We have followed [H1] for the proofs of Theorems 7.1 and 7.3. Proposition 7.2 is entirely elementary and Proposition 7.4 is a variant on the "Pietsch factorization" theme, which was first exploited in the C^*-context in [P5]. Lemma 7.6 and Proposition 7.11 come from [H1].

Shortly after [H1] appeared, Haagerup told me that his original proof of Theorem 7.5 used my result in [P5] but that he later found the argument published in [H1] which does not use it explicitly, together with the new proof of my result which is given in the appendix of [H1], and which we followed in the proof of Theorem 7.3. While based on the same ideas, the proof of Theorem 7.5 given in Chapter 7 is different from the one in [H1], but I suspect that it is actually very close to Haagerup's original unpublished argument. Note however that we only obtain $\|u\|_{cb} \le \|u\|^4$ while Haagerup in [H1] gets $\|u\|_{cb} \le \|u\|^3$.

As mentioned in the text, E. Christensen worked extensively on the derivation problem (see [C1, C2, C3]) and obtained Theorem 7.5 (in [C3]), but only in the irreducible case. Upon seeing Haagerup's proof, he understood how his argument in [C3] could be modified to also yield Theorem 7.5 in the cyclic case (see the addendum to [C3]).

We also refer the reader to [K2-3] for Kadison's viewpoint on the ideas around Theorem 7.5.

We should mention that in [VZ] an affirmative answer is given when A is arbitrary but under the additional assumption that the image $u(A)$ lies in a *finite* von Neumann algebra. (This result also was recently rediscovered by G.A. Robertson [Rob].)

Corollary 7.12 comes from [DCH]. Proposition 7.13 and its corollary come from [H1]. Lemma 7.17 is elementary and should be well known. Proposition 7.18 is implicit in [H1]. Lemma 7.19 is elementary.

The case of nuclear C^*-algebras (*i.e.* Theorem 7.16) was observed by Bunce [Bu] and independently by Christensen [C3].

Theorem 7.21 is due to Kirchberg [Ki]. (Theorem 7.20 also appears in [Ki] but is "folklore".)

In [C4], Christensen showed that every bounded unital homomorphism of a II_1 factor with property Γ is similar to a *-representation.

The most natural case of a C^*-algebra A for which the similarity Problem 0.2 for bounded unital homomorphisms $u: A \to B(H)$ is still open seems to be that of the reduced C^*-algebra $C^*_\lambda(F_2)$ of the free group with (say) 2 generators.

8. Completely bounded maps in the Banach space setting

Summary: In this chapter, we introduce the notion of p-complete boundedness for $1 \leq p < \infty$. The case $p = 2$ is the classical one, presented above in chapter 3. In this setting, we extend the fundamental factorization of c.b. maps. We show that the extension property still holds under suitable assumptions. Finally, we apply these results to the similarity problem for homomorphisms in this broader framework.

Let $1 \leq p < \infty$. One can also define a notion of complete boundedness adapted to the L_p-case. If X_1, Y_1 are Banach spaces and if $S \subset B(X_1, Y_1)$ is a subspace, we will say that an operator $u\colon S \to B(X, Y)$ is p-completely bounded (in short $p - c.b.$) if there is a constant C such that

$$\sup_n \|I_{B(\ell_p^n)} \otimes u\colon M_n(S) \to M_n(B(X,Y))\| \leq C$$

where $M_n(B(X,Y))$ and $M_n(S)$ are now equipped by the norms induced by

$$B(\ell_p^n(X), \ell_p^n(Y)) \quad \text{and} \quad B(\ell_p^n(X_1), \ell_p^n(Y_1)).$$

We denote by $\|u\|_{pcb}$ the smallest constant C for which this holds. Clearly, $\|u\| \leq \|u\|_{pcb}$.

There is a factorization of $p - c.b.$ maps analogous to Theorem 3.6 in the case $X_1 = Y_1 = L_p(\mu)$. Before we state this, we need a suitable extension of representations and compressions of representations to the L_p-setting. We first define a generalization of the Banach space valued L_p-spaces. Let I be an arbitrary set and let $\Lambda \subset \mathbb{R}^I$ be an order ideal, i.e. a linear subspace such that for $\psi, \varphi \in \mathbb{R}^I$ we have

$$(|\psi| \leq |\varphi| \quad \text{and} \quad \varphi \in \Lambda) \Rightarrow (\psi \in \Lambda).$$

Let $f\colon \Lambda \to \mathbb{R}$ be any positive linear form and let X be a Banach space. We wish to define the space of "X-valued p-integrable functions on I with respect to f." It turns out that we need to go beyond Bochner's theory of integration for X-valued measurable functions. We introduce the space $L(p, X)$ formed by all the functions $\psi\colon I \to X$ such that $\xi \to \|\psi(\xi)\|_X^p$ belongs to Λ and we define

$$N_p(\psi) = (f(\|\psi(\cdot)\|_X^p))^{1/p}.$$

Clearly N_p is a semi-norm on $L(p, X)$. After passing to the quotient by the kernel of N_p and completing we obtain a Banach space which we denote by $\Lambda_p(I, f; X)$ (or simply $\Lambda_p(X)$ if there is no risk of confusion).

If X is one dimensional, we denote the resulting space simply by $\Lambda_p(I, f)$. Note that this space is clearly a Banach lattice with a norm satisfying $\|x + y\|^p = \|x\|^p + \|y\|^p$ whenever $|x| \wedge |y| = 0$. Hence by a well known result of Kakutani, $\Lambda_p(I, f)$ is an L_p-space. More generally, whenever X is an L_p-space $\Lambda_p(I, f; X)$ also is an L_p-space. Now let X_1, Y_1 be Banach spaces. For any a in $B(X_1, Y_1)$, we denote by

$$\pi(a)\colon \Lambda_p(I, f; X_1) \longrightarrow \Lambda_p(I, f; Y_1)$$

the map which takes the equivalence class of $\xi \to \psi(\xi)$ to that of $\xi \to a\psi(\xi)$.

Let $E_2 \subset E_1 \subset \Lambda_p(I, f; X_1)$ and let $F_2 \subset F_1 \subset \Lambda_p(I, f; Y_1)$ be subspaces such that

$$(8.1) \qquad\qquad \pi(a)E_1 \subset F_1 \quad \text{and} \quad \pi(a)E_2 \subset F_2$$

for all a in S. Then (see the discussion at the begining of section 4) there is a natural "induced" map (recall $S \subset B(X_1, Y_1)$)

$$\hat{\pi}\colon S \longrightarrow B(E_1/E_2, F_1/F_2).$$

This map satisfies $\|\hat{\pi}\|_{pcb} \leq \|\pi_{|S}\|_{pcb} \leq 1$ (this is easy and left to the reader). We can now state the generalized version of Theorem 3.6.

Theorem 8.1. Let $S \subset B(X_1, Y_1)$ and let $u\colon S \to B(X, Y)$ be a $p - c.b.$ map. Then there are $I, f, E_2 \subset E_1, F_2 \subset F_1$ and $\hat{\pi}$ as above and there are operators

$$V_1\colon X \to E_1/E_2 \quad \text{and} \quad V_2\colon F_1/F_2 \to Y$$

with $\|V_1\| \, \|V_2\| \leq \|u\|_{pcb}$ such that

$$(8.2) \qquad\qquad \forall \, a \in S \qquad u(a) = V_2 \hat{\pi}(a) V_1.$$

Conversely, if such a factorization of u is possible then u is $p - c.b.$ and $\|u\|_{pcb} \leq \|V_1\| \, \|V_2\|$.

In particular we have

Corollary 8.2. The properties of Theorem 5.8 are equivalent to

(iv) φ defines a $p - c.b.$ Schur multiplier on $B(\ell_p(T), \ell_p(S))$ with $p - c.b.$ norm $\leq C$.

Proof: The fact that (ii) in Theorem 5.8 implies the above (iv) can be proved exactly as the corresponding implication (ii)$\Rightarrow$(iv) in the proof of Theorem 5.1. The converse is obvious. $\square$

To prove Theorem 8.1, we will proceed basically as in chapter 3. We let $I = B(X, X_1)$. For all ξ in I, all $z = \sum a_k \otimes x_k$ in $S \otimes X$ and all x in X, we denote

$$\xi.z = \sum a_k \xi(x_k)$$

$$\widehat{u}(z) = \sum u(a_k) x_k$$

and

$$\widehat{z}(\xi) = \xi.z, \quad \widehat{x}(\xi) = \xi(x).$$

For $z_i \in S \otimes X$ and $x_i \in X$ we write

$$(z_i) < (x_i)$$

if

$$\sum \|\xi.z_i\|_{Y_1}^p \leq \sum \|\xi.x_i\|_{X_1}^p, \text{ for all } \xi \text{ in } I.$$

For example, consider a matrix (a_{ij}) in the unit ball of $M_n(S)$ (i.e. such that $\|(a_{ij})\|_{\ell_p^n(X_1) \to \ell_p^n(Y_1)} \leq 1$). Let $x_1, \ldots, x_n$ be arbitrary in X and let $z_i = \sum_j a_{ij} \otimes x_j$. Then, it is very easy to verify that $(z_i) < (x_i)$.

We will now prove the following substitute for Lemma 3.3 by a different argument using convexity.

Lemma 8.3. *If X_1, Y_1, Y are arbitrary Banach spaces but if $X = \ell_1(\Gamma)$ then the map $u : S \to B(X, Y)$ is p-c.b. with $\|u\|_{pcb} \leq C$ iff*

(8.3)   *for all finite sequences (z_i) in $S \otimes X$ and all finite sequences (x_i) in X,*

$$(z_i) < (x_i) \Rightarrow \sum \|\widehat{u}(z_i)\|_Y^p \leq C^p \sum \|x_i\|_X^p.$$

Remark 8.4. Consider $x_1, \ldots, x_n$ in X. Assume that $z \in S \otimes X$ is such that if $\xi \in I$, $\xi(x_i) = 0$ for all $i \leq n$ implies $\xi.z = 0$. Then we claim that there are a_j in S such that $z = \sum a_j \otimes x_j$. This is a simple linear algebraic fact. Indeed, consider η in X^* such that $\eta \in \{x_1, \ldots, x_n\}^\perp$ and consider y arbitrary in X_1. Let $\xi = \eta \otimes y : X \to X_1$. Then by assumption we must have $\xi.z = 0$. Without loss of generality, we can assume that (x_j) are linearly independent. Let x_j^* be the functionals biorthogonal to (x_j) in X^*. Let $a_j = (I_S \otimes x_j^*)(z) \in S$. We claim that $z = \sum a_j \otimes x_j$ because $z' = z - \sum a_j \otimes x_j$ vanishes on any element of $X_1 \otimes X^*$. Indeed if η' in X^* and y in X_1 are arbitrary, then $\eta = \eta' - \sum \eta'(x_i) x_i^* \in \{x_1, \ldots, x_n\}^\perp$ hence by the first part of this remark if $\xi = \eta \otimes y$ we must have $\xi.z = 0$ or equivalently $< \eta' \otimes y, z' >= 0$ so that we conclude $z' = 0$.

The proof of Lemma 8.3 will use the following simple fact, which extends Lemma 5.2. We denote below by $(e_\gamma)_{\gamma \in \Gamma}$ the unit vector basis of $\ell_1(\Gamma)$.

Lemma 8.5. *Let $x_1, \ldots, x_n$ in $X = \ell_1(\Gamma)$ be such that $\sum \|x_i\|^p \leq 1$ and let X_1 be arbitrary. Then there are scalars λ_γ such that $\sum_{\gamma \in \Gamma} |\lambda_\gamma|^p \leq 1$ and an operator $B : \ell_p(\Gamma) \to \ell_p^n$ such that*

(8.4) $$\|B \otimes I_X\|_{\ell_p(\Gamma, X_1) \to \ell_p^n(X_1)} \leq 1$$

and such that the corresponding matrix $(b_{j\gamma})$ satisfies

$$x_j = \sum_{\gamma \in \Gamma} b_{j\gamma} \lambda_\gamma e_\gamma.$$

Proof: This is quite simple (as above for Lemma 5.2). We may assume $x_j \neq 0$. Let $x'_j = x_j(\|x_j\|^{-1})$ so that $\sum_\gamma |x'_j(\gamma)| = 1$. We have then for all ξ in $c_0(\Gamma, X_1)$ by convexity

$$\sum_j \| \sum_\gamma x_j(\gamma)\xi(\gamma)\|_{X_1}^p \leq \sum_j \|x_j\|^p \| \sum_\gamma x'_j(\gamma)\xi(\gamma)\|^p$$

$$\leq \sum_j \|x_j\|^p \sum_\gamma |x'_j(\gamma)| \|\xi(\gamma)\|^p$$

$$\leq \sum_\gamma \lambda_\gamma^p \|\xi(\gamma)\|^p$$

where we have set $\lambda_\gamma = (\sum_j \|x_j\|^p |x'_j(\gamma)|)^{1/p}$. Hence, there exists an operator $B: \ell_p(\Gamma) \to \ell_p^n$ with $\|B\| \leq 1$ satisfying (8.4) and moreover such that

$$\forall \xi \in c_0(\Gamma, X_1) \quad \sum_\gamma x_j(\gamma)\xi(\gamma) = \sum_\gamma b_{j\gamma} \lambda_\gamma \xi(\gamma).$$

In other words, $x_j = \sum_\gamma b_{j\gamma} \lambda_\gamma e_\gamma$. Since $\sum |\lambda_\gamma|^p \leq 1$ this completes the proof.

□

Proof of Lemma 8.3: Assume u $p-c.b.$ with $\|u\|_{pcb} \leq C$. Let $(z_i), (x_i)$ be as in (8.3) and assume $\sum \|x_i\|^p = 1$. Let $\lambda_\gamma, b_{j\gamma}$ be as in Lemma 8.5. Let $\bar{x}_\gamma = \lambda_\gamma e_\gamma$. Then $\sum \|\bar{x}_\gamma\|^p \leq 1$ and for all ξ in $B(X, X_1) \simeq \ell_\infty(\Gamma, X_1)$ we have

$$(8.5) \qquad \sum \|\xi.z_i\|^p \leq \sum \|\xi.x_i\|^p \leq \sum_\gamma \|\xi(\gamma)\lambda_\gamma\|^p.$$

By Remark 8.4, we have

$$z_i = \sum_j a_{ij} \otimes x_j \quad \text{with } x_j \in X, \ a_{ij} \in S,$$

hence (with an obvious abuse of notation)

$$z_i = \sum_\gamma \sum_j a_{ij} \otimes b_{j\gamma} \lambda_\gamma e_\gamma = \sum_\gamma c_{i\gamma} \otimes \lambda_\gamma e_\gamma$$

where

$$c_{i\gamma} = \sum_j a_{ij} b_{j\gamma} \in S.$$

We have by (8.5) for all ξ in $\ell_\infty(\Gamma, X_1)$

$$\sum_i \left\| \sum_\gamma c_{i\gamma} \lambda_\gamma \xi(e_\gamma) \right\|_{X_1}^p \leq \sum_\gamma |\lambda_\gamma|^p \|\xi(e_\gamma)\|^p,$$

hence (assuming $\lambda_\gamma \neq 0$ as we obviously may) the matrix $(c_{i\gamma})$ defines an element of the unit ball of $B(\ell_p(\Gamma, X_1), \ell_p^n(Y_1))$. Since we assume $\|u\|_{pcb} \leq C$, this implies that the matrix $(u(c_{i\gamma}))$ has norm at most C in $B(\ell_p(\Gamma, X), \ell_p^n(Y))$ hence

$$\sum_i \left\| \sum_\gamma u(c_{i\gamma})\lambda_\gamma e_\gamma \right\|^p \leq C^p \sum |\lambda_\gamma|^p$$

so that

$$\sum_i \|\widehat{u}(z_i)\|^p \leq C^p.$$

This shows the only if part. The converse is trivial.   □

We now describe the main point of the argument which is quite similar to the one used in chapter 3.

Let X, Y, X_1, Y_1 be arbitrary and asssume that u satisfies (8.3).
Let $I = B(X, X_1)$. Let $\Lambda \subset \mathbf{R}^I$ be the set of all functions $\phi : I \to \mathbf{R}$ for which there is a finite set $x_1, \ldots, x_n$ in X such that

$$\forall \xi \in I \qquad |\phi(\xi)| \leq \sum \|\widehat{x}_i(\xi)\|^p.$$

Then Λ is a sublattice of $\mathbf{R}^I$ equipped with the usual pointwise ordering. Actually Λ is an order ideal, this means that if $\phi, \psi \in \mathbf{R}^I, |\psi| \leq |\phi|$ and $\phi \in \Lambda$ then $\psi \in \Lambda$. As in the proof of Theorem 3.4, we define a sublinear functional

$$p(\phi) = \inf \left\{ C^p \sum \|x_i\|^p \,\middle|\, \phi \leq \sum \|\widehat{x}_i\|^p \right\}$$

and a superlinear functional on $\Lambda_+ = \Lambda \cap \mathbf{R}_+^I$

$$q(\phi) = \sup \left\{ \sum \|\widehat{u}(z_i)\|^p \,\middle|\, \sum \|\widehat{z}_i\|^p \leq \phi \right\}.$$

Since u is assumed to satisfy (8.3), we have $q \leq p$ on Λ_+, hence, by Corollary 3.2, we find a (necessarily positive) linear form $f : \Lambda \to \mathbf{R}$ such that $q(\phi) \leq f(\phi) \leq p(\phi)$ for all ϕ in Λ_+. In particular (with a slightly abusive but more suggestive notation) we obtain

(8.6) $\forall z \in S \otimes X \qquad \|\widehat{u}(z)\|_Y^p \leq f(\|\widehat{z}\|_{Y_1}^p)$

(8.7) $\forall x \in X \qquad f(\|\widehat{x}\|_{X_1}^p) \leq C^p \|x\|^p.$

Since the converse is obvious, we have proved that (8.3) holds iff there is an f as above such that (8.6) and (8.7) hold.

This leads us to consider the generalized space of X_1-valued L_p-functions which we denoted by $\Lambda_p(I, f; X_1)$ or more simply $\Lambda_p(X_1)$. These spaces were introduced at the beginning of this chapter, they go "beyond" the usual vector valued L_p-spaces in Bochner's sense.

Now let Y_1 be another Banach space. For any a in $B(X_1, Y_1)$ we have defined above $\pi(a) : \Lambda_p(X_1) \to \Lambda_p(Y_1)$ as the natural action of a on $\Lambda_p(X_1)$.

Recapitulating, we obtain the following result.

Theorem 8.6. *Let X, Y, X_1, Y_1 be arbitrary Banach spaces. Then, if u satisfies (8.3), there is a set I, an ideal $\Lambda \subset \mathbf{R}^I$, a linear functional $f : \Lambda \to \mathbf{R}$ and subspaces $M \subset \Lambda_p(I, f; X_1)$, $M' \subset \Lambda_p(I, f; Y_1)$ such that $\pi(a)M \subset M'$ for all a in S and there are operators $v_1 : X \to M$ and $v_2 : M' \to Y$ with $\|v_1\| \, \|v_2\| \leq C$ such that*

$$\forall a \in S, \quad u(a) = v_2 \pi(a) v_1.$$

Proof: Assume (8.3). Let f be the linear form obtained above satisfying (8.6) and (8.7). For any x in X we define $v_1 x$ as the element of $\Lambda_p(X_1)$ associated to $\hat{x}$. By (8.7) $\|v_1\| \leq C$. Let M be the closure of the image of v_1. Let $\mathcal{M}' \subset L(p, Y_1)$ be the set of all $\hat{z}$ for some z in $S \otimes X$. Let $\sigma : L(p, Y_1) \to \Lambda_p(Y_1)$ be the quotient map and let M' be the closure in $\Lambda_p(Y_1)$ of $\sigma(\mathcal{M}')$. By (8.6) we have an operator $v_2 : M' \to Y$ such that $\|v_2\| \leq 1$ and $\hat{u}(z) = v_2 \sigma \hat{z}$ for all z in $S \otimes X$. In particular $u(a)x = v_2 \pi(a) v_1 x$ for all x in X. This proves the announced implication. Note that actually the converse is also true but we leave this to the reader. □

Corollary 8.7. *In the same situation as in Theorem 8.1, if $X = \ell_1(\Gamma)$ and $Y = \ell_\infty(\Gamma')$ for some sets Γ and Γ' then the conclusion of Theorem 8.1 holds with $E_1 = \Lambda_p(I, f; X_1)$, $F_1 = \Lambda_p(I, f; Y_1)$ and $E_2 = \{0\}, F_2 = \{0\}$, so that we have $E_1/E_2 = \Lambda_p(I, f; X_1)$, $F_1/F_2 = \Lambda_p(I, f; Y_1)$ and $u(a) = V_2 \pi(a) V_1$ for all a in S.*

Proof: This is immediately deduced from Theorem 8.6 and Lemma 8.3, by the extension property of $Y = \ell_\infty(\Gamma')$. Indeed, with the notation in Theorem 8.6, the operator v_2 admits an extension $V_2 : \Lambda_p(Y_1) \to Y$ such that $\|V_2\| = \|v_2\|$. Let $V_1 : X \to \Lambda_p(X_1)$ be the operator v_1 considered as acting into $\Lambda_p(X_1)$. We have $u(a) = V_2 \pi(a) V_1$ for all a in S. □

Proof of Theorem 8.1. Consider now arbitrary spaces X, Y. It is well known that we can find sets Γ and Γ' and maps $Q : \ell_1(\Gamma) \to X$ and $J : Y \to \ell_\infty(\Gamma')$ such that J is an isometric embedding and Q is a metric surjection (i.e., Q^* is an isometric embedding).

Assume $\|u\|_{pcb} \leq C$. Then clearly, the map $a \to Ju(a)Q$ is $p - c.b.$ with $p - c.b.$ norm $\leq C$. Applying Corollary 8.7 to this map, we find f, $\Lambda_p(X_1)$, $\Lambda_p(Y_1)$, $\pi, w_1 : \ell_1(\Gamma) \to \Lambda_p(X_1)$ and $w_2 : \Lambda_p(Y_1) \to \ell_\infty(\Gamma')$ such that

$$\forall a \in S \qquad Ju(a)Q = w_2 \pi(a) w_1.$$

Let then $E_1 = \overline{w_1(\ell_1(\Gamma))}, E_2 = \overline{w_1(\ker Q)}, F_1 = w_2^{-1}(JY)$ and $F_2 = \ker w_2$. Clearly (8.1) holds and we can define unambiguously operators $V_1 : X \to E_1/E_2$ and $V_2 : F_1/F_2 \to Y$ by setting $\forall x \in X$, $V_1(x) = \sigma w_1 \check{x}$ where $\sigma : E_1 \to E_1/E_2$ denotes the quotient map and $\check{x}$ is any element of $Q^{-1}(\{x\})$. For all y in F_1/F_2 say $y = h + F_2$ for some h in F_1 we define $V_2 y = w_1 h$. It is easy to check that (8.2) holds and $\|V_1\| \, \|V_2\| \leq \|w_2\| \, \|w_1\| \leq C$, so that we obtain the first assertion in Theorem 8.1. The converse is obvious (since by the remarks preceding Theorem 8.1 we have $\|\hat{\pi}\|_{pcb} \leq 1$). □

Remark. If both X_1 and X are finite dimensional, we can obtain Theorem 8.1 and Theorem 8.6 with $L_p(\mu; X_1)$ and $L_p(\mu; Y_1)$ (for some measure space (Ω, μ)) in the place of $\Lambda_p(I, f; X_1)$ and $\Lambda_p(I, f; Y_1)$. Indeed, we can always in the preceding argument restrict I to be the unit ball of $B(X, X_1)$. Then I is a compact set and we can replace Λ by $\Lambda \cap C(I)$ (here $C(I)$ denotes the space of all continuous functions on I). Then the positive linear form f such that $f \leq p$ that we found above can be extended to a positive linear form on $C(I)$, therefore there is a positive measure μ on I such that

$$\forall x \in X \qquad \int \|\widehat{x}(\xi)\|^p \, d\mu(\xi) \leq C^p \|x\|^p$$

$$\forall a_k \in S \quad \forall x_k \in X \qquad \left\| \sum u(a_k) x_k \right\|^p \leq \int \left\| \sum a_k \xi(x_k) \right\|^p \, d\mu(\xi).$$

The functions $\widehat{x}$ and $\widehat{z}$ being continuous on I, the preceding argument yields finally $L_p(\mu; X_1)$ and $L_p(\mu; Y_1)$ instead of $\Lambda_p(X_1)$ and $\Lambda_p(Y_1)$.

Remark. It is easy to deduce from Theorem 8.1 that *in certain special cases* every $p - c.b.$ map defined on S has an extension to a $p - c.b.$ map defined in the whole of $B(X_1, Y_1)$. Indeed, assume that there are constants C_1 and C_2 satisfying the following

(8.8) Whenever M' is any subspace of any space $\Lambda_p(Y_1)$, every operator

$T_2 : M' \rightarrow Y$ extends to an operator $\widetilde{T}_2 : \Lambda_p(Y_1) \rightarrow Y$ with $\|\widetilde{T}_2\| \leq C_1 \|T_2\|$.

(8.9) For any subspace N of any space $\Lambda_p(X_1)$ every operator

$T_1 : X \rightarrow \Lambda_p(X_1)/N$, admits a lifting $\widetilde{T}_1 : X \rightarrow \Lambda_p(X_1)$ with $\|\widetilde{T}_1\| \leq C_2 \|T_1\|$.

Then we can state as an immediate consequence of Theorem 8.1.

Corollary 8.8. *In the situation of Theorem 8.1, with the assumptions (8.8) and (8.9), every $p - c.b.$ operator $u : S \rightarrow B(X, Y)$ admits a $p - c.b.$ extension $\widetilde{u} : B(X_1, Y_1) \rightarrow B(X, Y)$ satisfying*

$$\|\widetilde{u}\|_{pcb} \leq C_1 C_2 \|u\|_{pcb}.$$

Proof: We use the same notation as in Theorem 8.1. We apply (8.8) (resp. (8.9)) to the operator T_2 (resp. T_1) obtained by composing V_2 (resp. V_1) with the quotient mapping $F_1 \rightarrow F_1/F_2$ (resp. with the inclusion mapping $E_1/E_2 \rightarrow \Lambda_p(X_1)/E_2$). $\qquad\qquad\qquad\qquad\qquad\qquad\qquad\qquad\qquad\qquad\qquad\qquad\qquad$ □

Remark. Of course, (8.8) (resp. (8.9)) holds with $C_1 = 1$ (resp. $C_2 = 1 + \epsilon$, $\epsilon > 0$) if $Y = \ell_\infty(\Gamma')$ (resp. $X = \ell_1(\Gamma)$).

Remark. Moreover, if $p = 2$ and if Y_1 (resp. X_1) is a Hilbert space, then $\Lambda_2(Y_1)$ (resp. $\Lambda_2(X_1)$) also is a Hilbert space, so that (8.8) (resp. (8.9)) holds with $C_1 = 1$ (resp. $C_2 = 1$). Thus, we recover the extension property already obtained in Corollary 3.8.

Remark. More generally, by an extension theorem of Maurey [Ma], (8.8) happens if $2 \leq p < \infty$, with Y_1 of type 2 and Y of cotype 2. By duality Maurey's result shows that (8.9) will hold if $1 < p \leq 2$, with X_1^* type 2 and X^* cotype 2. However, we should point out that we only obtain a lifting going into the bidual of $\Lambda_p(X_1)$, but if we moreover assume that Y is reflexive, or is a dual space (or merely complemented in its bidual), then we can get rid of the biduals in the resulting factorization.

Now, returning to Theorem 8.1, assume $X_1 = Y_1$ and let $E_2 \subset E_1 \subset \Lambda_p(I, f; X_1)$ be subspaces invariant under $\pi(S)$ i.e. such that

$$\pi(a)E_1 \subset E_1 \quad \text{and} \quad \pi(a)E_2 \subset E_2$$

for all a in S. Then there is a natural "induced" map

$$(8.10) \qquad\qquad \hat{\pi}: S \to B(E_1/E_2)$$

which we call the compression of $\pi_{|S}$ to E_1/E_2. If S is a subalgebra of $B(X_1)$, then $\hat{\pi}$ is a homomorphism. Note that $\|\hat{\pi}\|_{pcb} \leq \|\pi\|_{pcb} \leq 1$.

Theorem 8.9. *Let A be a unital subalgebra of $B(X_1)$ and let $u: A \to B(X)$ be a unit preserving $p - c.b.$ homomorphism. Then there are I, f and $E_2 \subset E_1 \subset \Lambda_p(I, f; X_1)$ for which we can define the compression $\hat{\pi}$ as in (8.10) above and there is an isomorphism $S: X \to E_1/E_2$ with $\|S\| \, \|S^{-1}\| \leq \|u\|_{pcb}$ such that*

$$\forall a \in A \qquad u(a) = S^{-1}\hat{\pi}(a)S.$$

Conversely, if u has this form then we have

$$\|u\|_{pcb} \leq \|S\| \, \|S^{-1}\|.$$

Proof: We first apply Theorem 8.1. Let then $\mathcal{E}_1$ (resp. $\mathcal{E}_2$) be the closed span of $\{\pi(a)E_1 \mid a \in A\}$ (resp. of $\{\pi(a)E_2 \mid a \in A\}$). Then, since A is unital and $\pi(I) = I$, we have

$$E_1 \subset \mathcal{E}_1 \subset F_1 \quad \text{and} \quad E_2 \subset \mathcal{E}_2 \subset F_2.$$

Hence, we have natural maps

$$(8.11) \qquad\qquad E_1/E_2 \to \mathcal{E}_1/\mathcal{E}_2 \quad \text{and} \quad \mathcal{E}_1/\mathcal{E}_2 \to F_1/F_2.$$

Moreover, $\mathcal{E}_1$ and $\mathcal{E}_2$ are invariant under $\pi(A)$. Therefore, if we compose the maps V_1, V_2 from Theorem 8.1 with the maps (8.11), we can modify the conclusion of Theorem 8.1, in order to obtain it with $E_1 = \mathcal{E}_1 = F_1$ and $E_2 = \mathcal{E}_2 = F_2$. In that case, the induced mapping $\hat{\pi}$ is a homomorphism from A to $B(\mathcal{E}_1/\mathcal{E}_2)$. Therefore, we are in the situation to apply Proposition 4.2 (recall that since we assume $u(I) = I$, we must have $V_2 V_1 = I$ in Theorem 8.1) and this gives the first part of Theorem 8.9. The converse is obvious. $\qquad\qquad\square$

Finally, in the particular case when X_1 is an L_p-space, we have

Corollary 8.10. *Let (Ω, μ) be a measure space and X an arbitrary Banach space. Let A be a unital subalgebra of $B(L_p(\mu))$ and let $u: A \to B(X)$ be a unit preserving $p - c.b.$ homomorphism. Then there is a Banach space Z which is a subspace of a quotient of an L_p-space and an isomorphism $S: X \to Z$ such that $\|S\| \, \|S^{-1}\| \leq \|u\|_{pcb}$ and*

$$\forall a \in A \qquad \|Su(a)S^{-1}\| \leq \|a\|.$$

This applies for instance if A is the disc algebra $A(D)$ viewed as a subspace of $B(L_p(\mathbf{T}, m))$, with the inclusion $A(D) \to B(L_p(\mathbf{T}, m))$ mapping an element a in $A(D)$ to the operator of multiplication by a.

Thus in particular we obtain the following extension of Corollary 4.7 (and of Corollary 4.13).

Corollary 8.11. *Let $T \colon X \to X$ be an operator on a Banach space X and let C be a constant. Assume that for all n and all matrices (P_{ij}) with polynomial entries we have*

$$(8.12) \qquad \|(P_{ij}(T))\|_{B(\ell_p^n(X), \ell_p^n(X))} \le C \sup_{z \in \mathbf{T}} \|(P_{ij}(z))\|_{B(\ell_p^n, \ell_p^n)}.$$

Then there is a space Z isometric to a subspace of a quotient of an L_p-space and an isomorphism $S \colon X \to Z$ such that

$$\|S T S^{-1}\|_{B(Z,Z)} \le 1.$$

Remark. The case of "related" operators can be treated analogously, exactly as in Corollary 4.14 above. This shows that if (8.12) holds for all polynomials without constant term, then T is related to a contraction on a subspace of a quotient of L_p.

Notes and Remarks on Chapter 8

The notion of "p-complete boundedness" was introduced and studied in [P4], which is the main reference for this chapter. The latter contains in particular a comparison of the spaces $\Lambda_p(I, f; X)$ with ultraproducts of the ordinary (Bochner sense) $L_p(\mu; X)$ spaces. See also that paper for a more detailed discussion of 2-complete boundedness in connection with the notions of type 2 and cotype 2.

The notion of complete boundedness extends to the multilinear case, see [CS] for a survey of that. The method of [P4] can be extended to the bilinear case, but the n-linear case (for $n \geq 3$) remained open until C. Le Merdy settled it in [LeM1]. See also [LeM2-3] for related results.

References

AO   Akemann C. and Ostrand P. : Computing norms in group C^*-algebras. Amer.
        J. Math. 98 (1976) 1015-1047.

AS   Akcoglu M. A. and Sucheston L. : Dilations of positive contractions on L_p-
        spaces. Canad. Math. Bull. 20 (1977) 285-292.

A   Amir D. : Characterizations of inner product spaces. Birkhauser, 1986.

An1   Ando T. : On a pair of commutative contractions. Acta Sci. Math. 24 (1963)
        88-90.

An2   Ando T. : Unitary dilation for a triple of commuting contractions. Bull. Acad.
        Polon. Sci. 24 (1976), 851-853.

AFJS   Arias A., Figiel T., Johnson W. and Schechtman G. : Banach spaces which
        have the 2-summing property. Trans. Amer. Math. Soc. To appear.

Ar1   Arveson W. : Subalgebras of C^*-algebras. Acta Math. 123 (1969) 141-224.
        Part II. Acta Math. 128 (1972) 271–308.

Ar2   Arveson W. : Ten lectures on operator algebras. CBMS 55. Amer. Math. Soc.
        Providence (RI) 1984.

Be   Bennett G. : Schur multipliers. Duke Math J. 44 (1977) 603-639.

BGM   Bernard A. , Garnett J. and Marshall D. : Algebras generated by inner func-
        tions. J. Funct. Anal. 25 (1977) 275-285.

BP1   Blecher D. and Paulsen V. : Tensor products of operator spaces. J. Funct.
        Anal. 99 (1991) 262-292.

BP2   Blecher D. and Paulsen V. : Explicit construction of universal operator al-
        gebras and applications to polynomial factorization. Proc. Amer. Math. Soc.
        112 (1991) 839-850.

Ble   Blei R. : Multidimensional extensions of the Grothendieck inequality and
        applications. Arkiv för Mat. 17 (1979) 51-68.

Blo1   Blower G. : A multiplier characterization of analytic UMD spaces. Studia
        Math. 96 (1990) 117-124.

Blo2   Blower G. : On the complete polynomial bound of certain operators. Quar-
        terly J. Math. Oxford 43 (1992) 149-156.

Bo   Bourgain J. : On the similarity problem for polynomially bounded operators
        on Hilbert space. Israel J. Math. 54 (1986) 227-241.

B1   Bożejko M. : Remarks on Herz-Schur multipliers on free groups. Math. Ann.
        258 (1981) 11-15.

B2   Bożejko M. : Littlewood functions, Hankel multipliers and power bounded
        operators on a Hilbert space. Colloquium Math. 51 (1987) 35-42.

B3   Bożejko M. : Positive definite bounded matrices and a characterization of
        amenable groups. Proc. A.M.S. 95 (1985) 357-360.

B4   Bożejko M. : Positive-definite kernels, length functions on groups and a non-
        commutative von Neumann inequality. Studia Math. 95 (1989) 107-118.

B5   Bożejko M. : Uniformly bounded representations of free groups. J. für die
        Reine und Angewandte Math. 377 (1987) 170-186.

BF1   Bożejko M. and Fendler G. : Herz-Schur multipliers and completely bounded multipliers of the Fourier algebra of a locally compact group. Boll. Unione Mat. Ital. (6) 3-A (1984) 297-302.

BF2   Bożejko M. and Fendler G. : Herz-Schur multipliers and uniformly bounded representations of discrete groups. Arch. Math. 57 (1991) 290-298.

BR   Bratelli O. and Robinson D. : Operator algebras and quantum statistical mechanics II. Springer Verlag, New-York, 1981.

Bu   Bunce J. W. : The similarity problem for representations of C^*-algebras. Proc. Amer. Math. Soc. 81 (1981) 409-414.

CCFW   Carlson J., Clark D., Foias C. and Williams J. : Projective Hilbert $A(D)$-modules. Preprint.

C1   Christensen E. : Extensions of derivations. J. Funct. Anal. 27 (1978) 234-247.

C2   Christensen E. : Extensions of derivations II. Math. Scand. 50 (1982) 111-122.

C3   Christensen E. : On non self adjoint representations of operator algebras. Amer. J. Math. 103 (1981) 817-834.

C4   Christensen E. : Similarities of II_1 factors with property Γ. Journal Operator Theory 15 (1986) 281-288.

C5   Christensen E. : Perturbation of operator algebras II. Indiana Math. J. 26 (1977), 891–904.

CS   Christensen E. and Sinclair A. : A survey of completely bounded operators. Bull. London Math. Soc. 21 (1989) 417-448.

CRW   Coifman R., Rochberg R. and Weiss G. : Applications of transference: The L^p version of von Neumann's inequality and the Littlewood-Paley-Stein theory. Proc. Conf. Math. Res. Inst. Oberwolfach, Intern. ser. Numer. Math. vol. 40 (1978) 53-63, Birkhauser, Basel, 1978.

CLW   Cole B., Lewis K. and Wermer J. : Pick conditions on a uniform algebra and von Neumann inequalities. J. Funct. Anal. 107 (1992) 235-254.

CW   Cole B. and Wermer J. : Pick interpolation, von Neumann inequalities, and hyperconvex sets. Complex Potential Theory (edited by P. M. Gauthier). Nato ASI series (Proceedings, Montreal, July 26-august 6,1993) Kluwer Acad. Pub. Dordrecht, 1994.

Co   Connes A. : Classification of injective factors, Cases $II_1, II_\infty, III_\lambda, \lambda \neq 1$. Ann. Math. 104 (1976) 73-116.

Cow1   Cowling M. : Unitary and Uniformly bounded representations of some simple Lie groups. CIME Course 1980. Liguori, Napoli, 1982, p. 49-128.

Cow2   Cowling M. : Harmonic analysis on some nilpotent groups. Topics in modern harmonic Analysis Vol 1 I. N. D. A. M. Roma, 1983, p. 81-123.

Cow3   Cowling M. : Rigidity for lattices in semisimple Lie groups: von Neumann algebras and ergodic actions. Rend. Sem. Mat. Univers. Politecn. Torino 47 (1989) 1-37.

Cow4   Cowling M. : Sur les coefficients des représentations unitaires des groupes de Lie simples. Springer Lecture Notes in Math. 739 (1979) 132-178.

CoF   Cowling M. and Fendler G. : On representations in Banach spaces. Math. Ann. 266 (1984) 307-315.

CoH   Cowling M. and Haagerup U. : Completely bounded multipliers of the Fourier algebra of a simple Lie group of real rank one. Invent. Math. 96 (1989) 507-549.

CD   Crabb M. and Davie A. : von Neumann's inequality for Hilbert space operators. Bull. London Math. Soc. 7 (1975) 49-50.

Cu   Cuntz J. : Simple C^*-algebras generated by isometries. Comm. Math. Phys. 57 (1977) 173-185.

Dah   Daher M. : PhD Thesis, Université Paris 7, October 1993.

Dad   Davidson K. : Nest algebras. Pitman Research notes in Math. 191. Longman, London, New-York, 1988.

Dav   Davie A. M. : Quotient algebras of uniform algebras. J. London Math. Soc. 7 (1973) 31-40.

De   Dean D. : The equation $L(E, X^{**}) = L(E, X)^{**}$ and the principle of local reflexivity. Proc. Amer. Math. Soc. 40 (1973) 146-148.

DCH   de Cannière J. and Haagerup U. : Multipliers of the Fourier algebras of some simple Lie groups and their discrete subgroups. Amer. J. Math. 107 (1985) 455-500.

DF   Defant A. and Floret K. : Tensor norms and operator ideals. North-Holland, 1993.

DJT   Diestel J. , Jarchow H. and Tonge A. : Absolutely summing operators. Cambridge Univ. Press, 1995.

Di1   Dixmier J. : Les moyennes invariantes dans les semi-groupes et leurs applications. Acta Sci. Math. Szeged 12 (1950) 213-227.

Di2   Dixmier J. : les algèbres d'opérateurs dans l'espace Hilbertien (Algèbres de von Neumann). Gauthier-Villars, Paris, 1969. (Translated into: von Neumann algebras, North-Holland, 1981.)

Dix1   Dixon P. : The von Neumann inequality for polynomials of degree greater than two. J. London Math. Soc. 14 (1976) 369-375.

Dix2   Dixon P. : Q-algebras. Unpublished Lecture Notes. Sheffield University. 1975.

DD   Dixon P. and Drury S. : Unitary dilations, polynomial identities and the von Neumann inequality. Math. Proc. Cambridge Phil. Soc. 99 (1986) 115-122.

DP   Douglas V. and Paulsen V. : Hilbert modules over function algebras. Pitman Longman 1989.

Dr   Drury S. : Remarks on von Neumann's inequality. Banach spaces, Harmonic analysis, and probability theory. Proceedings (edited by R. Blei and S. Sydney), Storrs 80/81. Springer Lecture Notes 995, 14-32.

EM   Ehrenpreis L. and Mautner F. : Uniformly bounded representations of groups. Proc. Nat. Acad. Sc. 41 (1955) 231-233.

ER   Effros E. and Ruan Z.J. : A new approach to operator spaces. Canad. Math. Bull. 34 (1991) 329-337.

Ey   Eymard P. : L'algèbre de Fourier d'un groupe localement compact. Bull. Soc. Math. France 92 (1964) 181-236.

Fe1   Fendler G. : A uniformly bounded representation associated to a free set in a discrete group. Colloq. Math. 59 (1990) 223-229.

Fe2   Fendler G. : Herz-Schur multipliers and coefficients of bounded representations. Thesis Heidelberg.

FTN   Figa-Talamanca A. and Nebbia C. : Harmonic analysis and representation theory for groups acting on homogeneous trees. Cambridge Univ. Press, LMS Lecture notes series 162, Cambridge, 1990.

FTP   Figa-Talamanca A. and Picardello M. : Harmonic Analysis on Free groups. Marcel Dekker, New-York, 1983.

Fi   Fisher S. : The convex hull of the finite Blaschke products. Bull. A.M.S. 74 (1968) 1128-1129.

Fic   Ficken F.A. : Note on the existence of scalar products in normed linear spaces. Annals of Math. 45 (1944) 362-366.

Fo   Foguel S. : A counterexample to a problem of Sz.-Nagy. Proc. Amer. Math. Soc. 15 (1964) 788-790.

Foi   Foias C. : Sur certains théorèmes de J. von Neumann concernant les ensembles spectraux. Acta Sci. Math. 18 (1957) 15-20.

FW   Foias C. and Williams J. P. : On a class of polynomially bounded operators. Preprint (unpublished, approximately 1976).

Fou   Fournier J. : An interpolation problem for coefficients of H^∞ functions. Proc. Amer. Math. Soc. 42 (1974) 402-408.

Fug   Fuglede B. : A commutativity theorem for normal operators. Proc. Nat. Acad.
           Sci. U.S.A. 36 (1950) 35-40.

Gi1   Gilbert J. : L^p-convolution operators and tensor products of Banach spaces.
           Bull. Amer. Math. Soc. 80 (1974) 1127-1132.

Gi2   Gilbert J. : Convolution operators and Banach space tensor products, I, II,
           III. Unpublished preprints.

GL   Gilbert J. and Leih T. : Factorization, Tensor Products and Bilinear forms
           in Banach space Theory. In "Notes in Banach spaces" (edited by E. Lacey).
           University of Texas Press, Austin 1980.

Gr   Greenleaf F. : Invariant means on topological groups. Van Nostrand, New
           York 1969.

G   Grothendieck A. : Résumé de la théorie métrique des produits tensoriels topo-
           logiques. Boll. Soc. Mat. São-Paulo 8 (1956) 1-79.

Gu   Guralnick R. : A note on commuting pairs of matrices. Lin. and Multilin.
           Alg. 31 (1992) 71-75.

H1   Haagerup U. : Solution of the similarity problem for cyclic representations of
           C^*-algebras. Annals of Math. 118 (1983) 215-240.

H2   Haagerup U. : An example of a non-nuclear C^*-algebra which has the metric
           approximation property. Invent. Mat. 50 (1979) 279-293.

H3   Haagerup U. : Injectivity and decomposition of completely bounded maps. In
           "Operator algebras and their connection with topology and ergodic theory",
           Springer Lecture Notes in Math. 1132 (1985) 91-116.

H4   Haagerup U. : Decomposition of completely bounded maps on operator alge-
           bras. Unpublished manuscript. Sept.1980.

H5   Haagerup U. : A new upper bound for the complex Grothendieck constant.
           Israel J. Math. 60 (1987) 199-224.

H6   Haagerup U. : $M_0 A(G)$ functions which are not coefficients of uniformly
           bounded representations. Handwritten manuscript 1985.

HP1   Haagerup U. and Pisier G. : Linear operators between C^*-algebras. Duke
           Math. J. 71 (1993) 889-925.

HP2   Haagerup U. and Pisier G. : Factorization of analytic functions with values
           in non-commutative L_1-spaces. Canad. J. Math. 41 (1989) 882-906.

Had   Hadwin D. : Dilations and Hahn decompositions for linear maps. Canad. J.
           Math. 33 (1981) 826-839.

Ha1   Halmos P. : Ten problems in Hilbert space. Bull. Amer. Math. Soc. 76 (1970)
           887-933.

Ha2   Halmos P. : Introduction to Hilbert space. Chelsea. New-York, 1957.

HaN   Havin V. P. and Nikolski N. K. (editors) : Linear and complex analysis prob-
           lem book 3. Part 1. Lecture Notes in Math. 1573, Springer Verlag , Heidelberg,
           1994.

Hel   Helemskii A. Ya. : The homology of Banach and topological algebras. Kluwer
           Academic Publishers, Dordrecht, 1989.

Her1   Herz C. S. : Une généralisation de la notion de transformée de Fourier-
           Stieltjes. Ann. Inst. Fourier (Grenoble) 24 (1974) 145-157.

Her2   Herz C. S. : The theory of p-spaces with an application to convolution oper-
           ators. Trans. Amer. Math. Soc. 154 (1971) 69-82.

Ho1   Holbrook J. : Distortion coefficients for cryptocontractions. Lin. Alg. and
           Appl. 18 (1977) 229-256.

Ho2   Holbrook J. : Distortion coefficients for crypto-unitary operators. Lin. Alg.
           and Appl. 19 (1978) 189-205.

Ho3   Holbrook J. : Spectral dilations and polynomially bounded operators. Indiana
           Univ. Math. J. 20 (1971) 1027-1034.

Ho4   Holbrook J. : Polynomials in a matrix and its commutant. Linear Alg. Appl.
           48 (1982) 293-301.

Ho5   Holbrook J. : Inequalities of von Neumann type for small matrices. in "Function Spaces" (edited by K. Jarosz). Marcel Dekker, New-York, 1992.

Ho6   Holbrook J. : Interpenetration of ellipsoids and the polynomial bound of a matrix. Preprint, Sept. 1993.

J   Johnson B. E. : Cohomology in Banach algebras. Memoirs Amer. Math. Soc. 127 (1972).

Jo   Jolissaint P. : A characterization of completely bounded multipliers of Fourier algebras. Colloquium Math. 63 (1992) 311-313.

JoV   Jolissaint P. and Valette A. : Normes de Sobolev et convoluteurs bornés sur $L^2(G)$. Ann. Inst. Fourier 41 (1991) 797-822.

K1   Kadison R. : On the orthogonalization of operator representations. Amer. J. Math. 77 (1955) 600-620.

K2   Kadison R. : A note on the similarity problem. J. Operator Theory 26 (1991) 389-405.

K3   Kadison R. : On an inequality of Haagerup-Pisier. J. Op. Theory 29 (1993) 57-68.

K4   Kadison R. : Derivations of operator algebras. Ann. Math. 83 (1966) 280-293.

KR   Kadison R. and Ringrose J. : Fundamentals of the theory of operator algebras, vol. II. Academic Press. New-York, 1986.

Ka   Kahane J.P. : Some random series of functions. Heath Math. Monograph. 1968. New edition Cambridge Univ. Press, 1985.

Ke1   Kesten H. : Symmetric random walks on groups. Trans. Amer. Math. Soc. 92 (1959) 336-354.

Ke2   Kesten H. : Full Banach mean values on countable groups. Math. Scand. 7 (1959) 146-156.

Ki   Kirchberg E. : The derivation and the similarity problem are equivalent. Preprint. August 94.

Kö   König H. : On the complex Grothendieck constant in the n-dimensional case. Proc. of the Strobl Conf. Austria 1989, (edited by P. Mueller and W. Schachermayer) London Math. Soc. Lect. Notes 158 (1990) 181-198.

Kr1   Krivine J.L. : Constantes de Grothendieck sur les sphères et fonctions de type positif. Adv. in Math. 31 (1979) 16-30.

Kr2   Krivine J.L. : Sur la constante de Grothendieck. C. R. Acad. Sci. Paris Ser. A 284 (1977) 445-446.

KS   Kunze R.A. and Stein E. : Uniformly bounded representations and Harmonic Analysis of the 2×2 real unimodular group. Amer. J. Math. 82 (1960) 1-62.

Kw   Kwapień S. : On operators factorizable through L_p-space. Bull. Soc. Math. France Mémoire 31-32 (1972) 215-225.

Le   Lebow A. : A power bounded operator which is not polynomially bounded. Mich. Math. J. 15 (1968) 397-399.

L1   Leinert M. : Faltungsoperatoren auf gewissen diskreten Gruppen. Studia Math. 52, (1974) 149-158.

L2   Leinert M. : Abschtzung von Normen gewisser Matrizen und eine Anwendung. Math. Ann. 240 (1979) 13-19.

LeM1   Le Merdy C. : Factorizations of p-completely bounded multilinear maps. Preprint (1993). To appear in Pacific J. Math.

LeM2   Le Merdy C. : Representation of a quotient of a subalgebra of $B(X)$. Preprint (1994). To appear in Math. Proc. Camb. Phil. Soc.

LeM3   Le Merdy C. : Analytic factorizations and completely bounded maps. Israel J. Math. 88 (1994) 381-409.

LP   Lindalh L. and Poulsen F. : Thin sets in Harmonic Analysis. Marcel Dekker. New-York. 1971.

LiP   Lindenstrauss J. and Pelczynski A. : Absolutely summing operators in $\mathcal{L}_p$-spaces and their applications. Studia Math. 29 (1968) 275-326.

LT   Lindenstrauss J. and Tzafriri L. : Classical Banach spaces II. Springer Verlag Berlin, 1979.

LR   Lopez J. and Ross K. : Sidon sets. Marcel Dekker. New-York. 1975.

Los   Losert V. : Properties of the Fourier algebra that are equivalent to amenability. Proc. Amer. Math. Soc. 92 (1984) 347-354.

Lo   Lotto B. : Von Neumann's inequality for commuting, diagonalizable contractions, I. Proc. Amer. Math. Soc. To appear.

LoS   Lotto B. and Steger T. : Same title II. To appear.

LuP   Lust-Piquard F. : Opérateurs de Hankel 1-sommants de $\ell^1(\mathbb{N})$ dans $\ell^\infty(\mathbb{N})$ et multiplicateurs de $H^1(T)$. Comptes Rendus Acad. Sci. Paris. 299 (1984) 915-918.

LuPP   Lust-Piquard F. and Pisier G. : Non commutative Khintchine and Paley inequalities. Arkiv fr Mat. 29 (1991) 241-260.

MZ1   Mantero A.M. and Zappa A. : The Poisson transform on free groups and uniformly bounded representations. J. Funct. Anal. 51 (1983) 372-399.

MZ2   Mantero A.M. and Zappa A. : Uniformly bounded representations and L_p-convolution theorems on a free group. Harmonic Analysis (Proc. Cortona 1982) Springer Lecture Notes 992 (1983) 333-343.

M   Mascioni V. : Ideals of the disc algebra, operators related to Hilbert space contractions, and complete boundedness. Houston J. Math. 20 (1994) 299-311.

Mat   Mathieu M. : The cb-norm of a derivation. In "Algebraic Methods in Operator Theory" (edited by R. Curto and P. Jørgensen) Birkhauser, Boston, 1994, pp. 144–152.

Ma   Maurey B. : Un théorème de prolongement. C. R. Acad. Sci. Paris A 279 (1974) 329-332.

Mi   Misra G. : Curvature inequalities and extremal problems of bundle shifts. J. Operator Theory 11 (1984) 305-318.

MS1   Misra G. and Sastry S. : Completely contractive modules and associated extremal problems. J. Funct. Anal. 91 (1990) 213-220.

MS2   Misra G. and Sastry S. : Bounded modules, extremal problems and a curvature inequality. J. Funct. Anal. 88 (1990) 118-134.

Ml   Mlak W. : Algebraic polynomially bounded operators. Ann. Pol. Math. 29 (1974) 133-139.

Na   Nakazi T. : Commuting dilations and uniform algebras. Canad. J. Math. 42 (1990) 776-789.

Ne   Nebbia C. : Multipliers and asymptotic behaviour of the Fourier algebra of non amenable groups. Proc. Amer. Math. Soc. 84 (1982) 549-554.

Nel   Nelson E. : The distinguished boundary of the unit operator ball. Proc. Amer. Math. Soc. 12 (1961) 994-995.

vN   von Neumann J. : Eine spektraltheorie für allgemeine operatoren eines unitären raumes. Math. Nachr. 4 (1951) 258-281.

Ni   Nikolskii N. : Treatise on the shift operator. Springer Verlag, Berlin 1986.

Pag   Page L. : Bounded and compact vectorial Hankel operators. Trans. Amer. Math. Soc. 150 (1970) 529-540.

Par1   Parrott S. : Unitary dilations for commuting contractions. Pacific J. Math. 34 (1970) 481-490.

Par2   Parrott S. : On a quotient norm and the Sz.-Nagy-Foias lifting theorem. J. Funct. Anal. 30 (1978) 311-328.

Pat   Paterson A. : Amenability. A.M.S. Math. Surveys 29 (1988).

Pa1   Paulsen V. : Completely bounded maps and dilations. Pitman Research Notes in Math. 146, Longman, Wiley, New York, 1986.

Pa2   Paulsen V. : Completely bounded maps on C^*-algebras and invariant operator ranges. Proc. Amer. Math. Soc. 86 (1982) 91-96.

Pa3   Paulsen V. : Every completely polynomially bounded operator is similar to a contraction. J. Funct. Anal. 55 (1984) 1-17.

Pa4   Paulsen V. : Completely bounded homomorphisms of operator algebras. Proc. Amer. Math. Soc. 92 (1984) 225-228.

Pa5   Paulsen V. : Representations of function algebras, abstract operator spaces and Banach space geometry. J. Funct. Anal. 109 (1992) 113-129.

Pa6   Paulsen V. : Completely bounded maps on C^*-algebras and invariant operator ranges. Proc. Amer. Math. Soc. 86 (1982) 91-96.

PPP   Paulsen V. , Pearcy C. and Petrovic S. : On centered and weakly centered operators. J. Funct. Anal. 128 (1995) 87-101.

Ped   Pedersen G. : C^*-algebras and their automorphism groups. Academic Press, London, 1979.

Pe1   Peller V. : Estimates of functions of power bounded operators on Hilbert space. J. Oper. Theory 7 (1982) 341-372.

Pe2   Peller V. : An analogue of an inequality of J. von Neumann, isometric dilation of contractions, and approximation by isometries in spaces of measurable functions. Trudy Inst. Steklov, 155 (1981) 103-150, English translation in Proc. Steklov Inst. Math. (1983) 101-145.

Pe3   Peller V. : Analog of J. von Neumann's inequality for L_p space. Dokl. Akad. Nauk SSSR 231 (1976) 539-542. (Russian)

Pet   Petrovic S. : A dilation Theory for Polynomially Bounded Operators. J. Funct. Anal. 108 (1992) 458-469.

Pi   Pier J.P. : Amenable locally compact groups. Wiley, Interscience, New York 1984.

Pie   Pietsch A. : Operator ideals. North-Holland Amsterdam 1978.

P1   Pisier G. : Factorization of linear operators and the Geometry of Banach spaces. CBMS (Regional conferences of the A.M.S.) no 60, (1986) Reprinted with corrections 1987.

P2   Pisier G. : Factorization of operator valued analytic functions. Advances in Math. 93 (1992) 61-125.

P3   Pisier G. : Multipliers and lacunary sets in non-amenable groups. Amer. J. Math. 117 (1995) 337-376.

P4   Pisier G. : Completely bounded maps between sets of Banach space operators. Indiana Univ. Math. J. 39 (1990) 251-277.

P5   Pisier G. : Grothendieck's theorem for noncommutative C^*-algebras with an appendix on Grothendieck's constants. J. Funct. Anal. 29 (1978) 397-415.

P6   Pisier G. : Complex interpolation and regular operators between Banach lattices. Arch. Math. (Basel) 62 (1994) 261-269.

P7   Pisier G. : Non-commutative vector valued L_p-spaces and completely p-summing maps. Preprint To appear.

Po1   Popescu G. : Von Neumann inequality for $(B(H)^n)_1$. Math. Scand. 68 (1991) 292-304.

Po2   Popescu G. : Non-commutative disc algebras and their representations. 1994, Preprint to appear.

Po3   Popescu G. : Positive-definite functions on free semigroups. Preprint, 1994, to appear.

Po4   Popescu G. : Multi-analytic operators on Fock space. Math. Ann. to appear.

PyS   Pytlik T. and Szwarc R. : An analytic family of uniformly bounded representations of free groups. Acta Math. 157 (1986) 287-309.

Ri   Ringrose J. : Cohomology theory for operator algebras. Proc. Symp. Pure Math. 38 (1982) 229-252.

Rob   Robertson A.G. : Uniformly bounded group representations into a finite von Neumann algebra. Preprint 1993.

RR   Rosenblum M. and Rovnyak J. : Hardy classes and operator theory. Oxford
        Univ. Press, New York, 1985.

Ro   Rota G.C. : On models for linear operators. Comm. Pure Appl. Math. 13
        (1960) 468-472.

R1   Rudin W. : Fourier analysis on groups. Interscience. New York, 1962.

R2   Rudin W. : Convex combinations of unimodular functions. Bull. A.M.S. 75
        (1969) 795-797.

R3   Rudin W. : Functional Analysis. McGraw-Hill, New-York 1973.

R4   Rudin W. : Real and Complex Analysis. Third edition, Mc Graw-Hill, New-
        York, 1987.

S   Sakai S. : Derivations of W^*-algebras. Ann. Math. 83 (1966) 273-279.

Sa   Sarason D. : Generalized interpolation in H^∞. Trans. Amer. Math. Soc. 127
        (1967) 179-203.

SS   Sinclair A. and Smith R. : Hochschild cohomology of von Neumann algebras.
        LMS Lecture notes series. Cambridge Univ. Press, 1994.

Sm   Smith R. : Completely bounded maps between C^*-algebras. J. London Math.
        Soc. 27 (1983) 157-166.

St   Stampfli J. : The norm of a derivation. Pacific J. Math. 33 (1970) 737-747.

Sti   Stinespring W. : Positive functions on C^*-algebras. Proc. Amer. Math. Soc. 6
        (1966) 211-216.

SN   Sz.-Nagy B. : Completely continuous operators with uniformly bounded iter-
        ates. Publ. Math. Inst. Hungarian Acad. Sci. 4 (1959) 89-92.

SNF   Sz.-Nagy B. and Foias C. Harmonic analysis of operators on Hilbert space.
        Akademiai Kiadó, Budapest 1970.

TJ   Tomczak-Jaegermann N. : On the Rademacher averages and the moduli of
        convexity and smoothness of the Schatten classes S_p. Studia Math. 50 (1974)
        163-182.

To 1   Tonge A. : The von Neumann inequality for polynomials in several Hilbert
        Schmidt operators. J. London Math. Soc. 18 (1978) 519-526.

To 2   Tonge A. : Polarisation and the 2 dimensional Grothendieck inequality. Proc.
        Cambridge Phil. Soc. 95 (1984) 313-318.

Va1   Valette A. : Les représentations uniformément bornées associées à un arbre
        réel. Bull. Soc. Math. Belgique 42 (1990) 747-760.

Va2   Valette A. : Cocycles d'arbres et représentations uniformément bornées. C.
        R. Acad. Sci. Paris 310 (1990) 703-708.

V1   Varopoulos N. : On an inequality of von Neumann and an application of the
        metric theory of tensor products to Operators Theory. J. Funct. Anal. 16
        (1974) 83-100.

V2   Varopoulos N. : Tensor algebras over discrete spaces. J. Funct. Anal. 3 (1969)
        321-335.

V3   Varopoulos N. : On a commuting family of contractions on a Hilbert space.
        Rev. Roumaine Math. Pures Appl. 21 (1976) 1283-1285.

VZ   Vasilescu F.H. and Zsido L. : Uniformly bounded groups in finite W^*-algebras.
        Acta Sci. Math. 36 (1974) 189-192.

Wi1   Wittstock G. : Ein operatorwertigen Hahn-Banach Satz. J. Funct. Anal. 40
        (1981) 127–150.

Wi2   Wittstock G. : On matrix order and convexity. Functional analysis: surveys
        and recent results. Math. Studies 90, p. 175-188. North Holland, Amsterdam,
        1984.

W1   Wysoczanski J. : Characterization of amenable groups and the Littlewood
        functions on free groups. Colloquium Math. 55 (1988) 261-265.

W2   Wysoczanski J. : An analytic family of uniformly bounded representations of
        a free product of discrete groups. Pacific J. Math. 157 (1993) 373-385.

W3   Wysoczanski J. : Radial Herz-Schur multipliers on free products of discrete
          groups. Journal Funct. Anal. 129 (1995) 268-292.
Zs   Zsidó L. : The norm of a derivation in a W^*-algebra. Proc. Amer. Math. Soc.
          38 (1973) 147-150.

Subject Index

Notation Index

Lecture Notes in Mathematics

For information about Vols. 1–1439
please contact your bookseller or Springer-Verlag

Vol. 1479: S. Bloch, I. Dolgachev, W. Fulton (Eds.), Algebraic Geometry. Proceedings, 1989. VII, 300 pages. 1991.

Vol. 1480: F. Dumortier, R. Roussarie, J. Sotomayor, H. Żołądek, Bifurcations of Planar Vector Fields: Nilpotent Singularities and Abelian Integrals. VIII, 226 pages. 1991.

Vol. 1481: D. Ferus, U. Pinkall, U. Simon, B. Wegner (Eds.), Global Differential Geometry and Global Analysis. Proceedings, 1991. VIII, 283 pages. 1991.

Vol. 1482: J. Chabrowski, The Dirichlet Problem with L^2-Boundary Data for Elliptic Linear Equations. VI, 173 pages. 1991.

Vol. 1483: E. Reithmeier, Periodic Solutions of Nonlinear Dynamical Systems. VI, 171 pages. 1991.

Vol. 1484: H. Delfs, Homology of Locally Semialgebraic Spaces. IX, 136 pages. 1991.

Vol. 1485: J. Azéma, P. A. Meyer, M. Yor (Eds.), Séminaire de Probabilités XXV. VIII, 440 pages. 1991.

Vol. 1486: L. Arnold, H. Crauel, J.-P. Eckmann (Eds.), Lyapunov Exponents. Proceedings, 1990. VIII, 365 pages. 1991.

Vol. 1487: E. Freitag, Singular Modular Forms and Theta Relations. VI, 172 pages. 1991.

Vol. 1488: A. Carboni, M. C. Pedicchio, G. Rosolini (Eds.), Category Theory. Proceedings, 1990. VII, 494 pages. 1991.

Vol. 1489: A. Mielke, Hamiltonian and Lagrangian Flows on Center Manifolds. X, 140 pages. 1991.

Vol. 1490: K. Metsch, Linear Spaces with Few Lines. XIII, 196 pages. 1991.

Vol. 1491: E. Lluis-Puebla, J.-L. Loday, H. Gillet, C. Soulé, V. Snaith, Higher Algebraic K-Theory: an overview. IX, 164 pages. 1992.

Vol. 1492: K. R. Wicks, Fractals and Hyperspaces. VIII, 168 pages. 1991.

Vol. 1493: E. Benoît (Ed.), Dynamic Bifurcations. Proceedings, Luminy 1990. VII, 219 pages. 1991.

Vol. 1494: M.-T. Cheng, X.-W. Zhou, D.-G. Deng (Eds.), Harmonic Analysis. Proceedings, 1988. IX, 226 pages. 1991.

Vol. 1495: J. M. Bony, G. Grubb, L. Hörmander, H. Komatsu, J. Sjöstrand, Microlocal Analysis and Applications. Montecatini Terme, 1989. Editors: L. Cattabriga, L. Rodino. VII, 349 pages. 1991.

Vol. 1496: C. Foias, B. Francis, J. W. Helton, H. Kwakernaak, J. B. Pearson, H∞-Control Theory. Como, 1990. Editors: E. Mosca, L. Pandolfi. VII, 336 pages. 1991.

Vol. 1497: G. T. Herman, A. K. Louis, F. Natterer (Eds.), Mathematical Methods in Tomography. Proceedings 1990. X, 268 pages. 1991.

Vol. 1498: R. Lang, Spectral Theory of Random Schrödinger Operators. X, 125 pages. 1991.

Vol. 1499: K. Taira, Boundary Value Problems and Markov Processes. IX, 132 pages. 1991.

Vol. 1500: J.-P. Serre, Lie Algebras and Lie Groups. VII, 168 pages. 1992.

Vol. 1501: A. De Masi, E. Presutti, Mathematical Methods for Hydrodynamic Limits. IX, 196 pages. 1991.

Vol. 1502: C. Simpson, Asymptotic Behavior of Monodromy. V, 139 pages. 1991.

Vol. 1503: S. Shokranian, The Selberg-Arthur Trace Formula (Lectures by J. Arthur). VII, 97 pages. 1991.

Vol. 1504: J. Cheeger, M. Gromov, C. Okonek, P. Pansu, Geometric Topology: Recent Developments. Editors: P. de Bartolomeis, F. Tricerri. VII, 197 pages. 1991.

Vol. 1505: K. Kajitani, T. Nishitani, The Hyperbolic Cauchy Problem. VII, 168 pages. 1991.

Vol. 1506: A. Buium, Differential Algebraic Groups of Finite Dimension. XV, 145 pages. 1992.

Vol. 1507: K. Hulek, T. Peternell, M. Schneider, F.-O. Schreyer (Eds.), Complex Algebraic Varieties. Proceedings, 1990. VII, 179 pages. 1992.

Vol. 1508: M. Vuorinen (Ed.), Quasiconformal Space Mappings. A Collection of Surveys 1960-1990. IX, 148 pages. 1992.

Vol. 1509: J. Aguadé, M. Castellet, F. R. Cohen (Eds.), Algebraic Topology - Homotopy and Group Cohomology. Proceedings, 1990. X, 330 pages. 1992.

Vol. 1510: P. P. Kulish (Ed.), Quantum Groups. Proceedings, 1990. XII, 398 pages. 1992.

Vol. 1511: B. S. Yadav, D. Singh (Eds.), Functional Analysis and Operator Theory. Proceedings, 1990. VIII, 223 pages. 1992.

Vol. 1512: L. M. Adleman, M.-D. A. Huang, Primality Testing and Abelian Varieties Over Finite Fields. VII, 142 pages. 1992.

Vol. 1513: L. S. Block, W. A. Coppel, Dynamics in One Dimension. VIII, 249 pages. 1992.

Vol. 1514: U. Krengel, K. Richter, V. Warstat (Eds.), Ergodic Theory and Related Topics III, Proceedings, 1990. VIII, 236 pages. 1992.

Vol. 1515: E. Ballico, F. Catanese, C. Ciliberto (Eds.), Classification of Irregular Varieties. Proceedings, 1990. VII, 149 pages. 1992.

Vol. 1516: R. A. Lorentz, Multivariate Birkhoff Interpolation. IX, 192 pages. 1992.

Vol. 1517: K. Keimel, W. Roth, Ordered Cones and Approximation. VI, 134 pages. 1992.

Vol. 1518: H. Stichtenoth, M. A. Tsfasman (Eds.), Coding Theory and Algebraic Geometry. Proceedings, 1991. VIII, 223 pages. 1992.

Vol. 1519: M. W. Short, The Primitive Soluble Permutation Groups of Degree less than 256. IX, 145 pages. 1992.

Vol. 1520: Yu. G. Borisovich, Yu. E. Gliklikh (Eds.), Global Analysis – Studies and Applications V. VII, 284 pages. 1992.

Vol. 1521: S. Busenberg, B. Forte, H. K. Kuiken, Mathematical Modelling of Industrial Process. Bari, 1990. Editors: V. Capasso, A. Fasano. VII, 162 pages. 1992.

Vol. 1522: J.-M. Delort, F. B. I. Transformation. VII, 101 pages. 1992.

Vol. 1523: W. Xue, Rings with Morita Duality. X, 168 pages. 1992.

Vol. 1524: M. Coste, L. Mahé, M.-F. Roy (Eds.), Real Algebraic Geometry. Proceedings, 1991. VIII, 418 pages. 1992.

Vol. 1525: C. Casacuberta, M. Castellet (Eds.), Mathematical Research Today and Tomorrow. VII, 112 pages. 1992.